Living With Wildlife

Marilyn and Ron Leys

Published by

700 E. State Street • Iola, WI 54990-0001
Telephone: 715/445-2214

Please call or write for our free catalog or visit our website.
Our toll-free number to place an order or obtain a free catalog is 800-258-0929
or please use our regular business telephone 715-445-2214
for editorial comment and further information.
www.krause.com

Library of Congress: 00-101580
ISBN: 0-87341-857-3

Printed in the United States of America

TABLE OF CONTENTS

ACKNOWLEDGMENTS

I cast my line into thin air and reeled in the connections to much of the material that you will find in this book. My fishing hole was the Internet; my guide was Jon Margerum-Leys, in whose University of Michigan classroom I learned enough to fish confidently on my own. It was a worthwhile role reversal for a mother who is, herself, a teacher. The Internet led me to natural resources departments and university extension services in 19 states. These provided me with a wealth of books, booklets and brochures, as well as personal communications by telephone, e-mail and old-fashioned mail. I owe a particular debt to several Wisconsin DNR employees, including Gary Harden, who spent a lot of time filling me in on his programs, Paul Brandt, who started me out with a stack of printed material, Cindi Kohles, who gave me a guided tour of properties in my neighborhood that are under attack by invasive plants, and Kelly Kearns, who kindly answered a number of my questions about those invaders.

With only hope on my side, I sent out blind requests through the websites of The National Wildlife Federation, Wild Ones and several other organizations. One response led to another, resulting in some of the interviews in Chapter 7. To all of the people I interviewed, who welcomed a complete stranger into their gardens and fields, and to the others I interviewed for this book, I owe much thanks.

Sandra Stark and Julie Margerum-Leys spent many hours taking photos for the book. John Fore entrusted me with an extensive collection of photos snapped by the motion-sensitive cameras his company manufactures. Gary Eldred let me take my pick of slides of area prairies. Patricia Gilbert spent hours drawing, conferring and re-drawing in order to illustrate some of the ideas you will find here. And Lisa Ashley used her considerable knowledge to plan a garden for me and for the book.

For his ongoing help, I owe thanks to my husband, Ron, who kept me on track. The contacts he made during years of outdoor reporting led to this book, then opened doors to interviews that might not otherwise have occurred.

— Marilyn Leys

FOREWORD

Late one October afternoon, a fastidious chickadee perched at the top of a tall weed that was growing close to the house window where I stood watching it. The bird spent a good five minutes twisting itself into the most amazing postures as it delicately pecked at an assortment of seed heads.

That winter, after the first snow fell, hordes of goldfinches and house finches descended upon the tube feeders at our living room window, and juncos flocked to the ground beneath. On several occasions, a male cardinal brightened the day, first working up his courage in the upper branches of the maple tree across the drive from the feeders, now flitting down for the briefest of meals.

And on a sunny, unseasonably warm February afternoon, I nearly ran over our resident pheasant.

All of these encounters happened because of something I discovered while researching and writing this book.

In previous years, my job was to keep the fences bordering our lawn weed free. But working on the chart about birds' food preferences – helped along by my natural laziness – resulted in my leaving the weeds that the acrobatic chickadee and other birds discovered.

The renewed popularity of our bird feeders was the result of making an equally simple change. We had experienced several years of good luck with our feeders before a winter when nobody came to visit. By the time Ron hauled out the feeders the following fall, I had read a lot in preparation for writing. Let me try washing out the feeders before you fill them, I suggested. That won't help, Ron said. But I washed them anyway. Then he filled them with new seed – thistle seed for the goldfinches and black oil sunflower seed for everybody. The cleaning helped big-time – some days those feeders had to be filled twice. Maintaining the pheasant was a bit more complicated. He'd shown up for the first time the previous spring, announcing his presence with a piercing cackle. Then, all summer long, he'd turn up every now and then. He was such a beautiful gift that I wanted to keep seeing him, but my research suggested that a food plot was the way to do it and he'd arrived well after food-plot-planting season.

One day in early October, harvest season, I was strolling down the gravel road that leads away from our farm, passing between one neighbor's cornfield and another's hayfield before it reaches the state highway. In the distance I could hear the sound of our neighbor's tractor starting up. Suddenly it occurred to me that a food plot didn't need to start out life as a patch of corn planted solely for the enjoyment of wildlife – a food plot merely had to end up that way. Which is how we ended up paying our neighbor to leave a little of his corn standing. As it turned out, this practical farmer thought it was a wonderful idea to encourage the pheasant. He did, however, charge us full price for his crop. When snow fell into the remnant cornfield, it became clear that deer were also taking advantage of the corn, a side benefit that the family deer hunter did not mind at all. Then came the late winter day when the pheasant strolled within six inches of my car's tires, proof that the food plot was doing what I'd hoped for after all.

You might not own enough land to attract your own pheasant, but no matter how small your property, there is probably something you can add to entice whatever wildlife already lives nearby. Whether you have the time and enthusiasm to do something as complicated as installing a prairie or a stand of aspen or as simple as letting a few weeds hang around, you'll find many suggestions in this book – probably even more than you and your land can handle.

DEDICATION

In memory of Aldo Leopold, who taught us about the interrelationships of Man, animals, and the Earth we all share.

Chapter 1

EXPANDING OUR UNIVERSE

We live our day-to-day lives in a version of reality, of home, family, work, recreation, eating, sleeping. That's our universe.

But many of us know there is a deeper, more profound reality. Another universe. A universe that exists all around us, a universe that is parallel to ours, a universe we can see but that we can only occasionally enter.

Our kind once belonged to that other universe, the universe of the wild ones. We left it for the comforts of warm houses, food from the supermarket, flush toilets and sheets on our beds.

Something in many of us, however, yearns for a return to that universe of the wild ones, even if only as a spectator. And so we look out the office window at a gull flying free down the river, we look skyward as we leave the factory for the day, hoping to see a V of geese heading north, we slow the car on the way home to let a squirrel cross in front of us.

That gull, those geese and that squirrel each live part of their lives in our universe, though they never come to stay. But just seeing them, just knowing they are there, living their lives by their own rules, gladdens our hearts in a way that no zoo animals, no chickens in a pen, no cattle in a barn, no dog at the hearth ever can.

There are ways to invite those citizens of the wild universe to come closer, to make themselves at home in our back yards, around the lake cottage, in that 40 acres on which we hope to build a cabin some day.

Even though we will never be part of their lives, they can, in a profound way, become part of ours.

In summer, only a few feeders compete with the bird food that grows on shrubs near the Rolling Ground tavern. In winter, as many as eight feeders provide amusement for the tavern's customers.

There's a country tavern up the road from our place, Murphy's Rolling Ground, where Irish farmers gather to tell stories and share a laugh and a drink. It's nothing more, but also nothing less, than a country tavern. You'd feel comfortable there.

Big windows behind the bar overlook the hills that give this memory of a settlement its name. In front of the windows are six or eight bird feeders, kept ever-filled by Bonnie and Ron Murphy. Woods and fields just across the road are home to many species of birds, and evergreens in front of the tavern give the birds courage to sneak over for and snitch a seed.

Many of the bib-overalled farmers and electricians and well-drillers and such take a keen interest in those visitors from the wild universe, marking the change of the season in the late-winter yellowing of the goldfinches, noting the different woodpeckers that stop by and laughing aloud at the belligerent little hummingbirds that fight over places at the sugar-water tubes.

The wild birds are as much a part of the Rolling Ground tradition as the corned beef and cabbage at the annual St. Pat's dinner or the horses in the mini-parade at the 4th of July hog roast.

Marilyn and I live in the country, but these days not everyone who lives in the country is a farmer. Not by a long shot. Even in this 100-percent rural neighborhood in the southwest corner of Wisconsin, a very large percentage of people work over in town or do construction work or run computer-based operations out of their homes or are retired from city jobs.

They live on residential lots, not so different from those in a thousand suburbs across America, and in many ways not so different from back yards in America's cities.

Although I live on a farm today, raising beef cattle for laughs and a few bucks, most of my 63 years have been spent in closer contact with other people, from life in big and small cities, to close-in and farther-out suburbs, to military barracks.

I don't remember a time when that wild universe was not also part of my universe.

No matter what the calendar said, spring was when the robins arrived to patrol our city yard for worms and to build their nests under the eaves of the garage and to sing their evening songs from the telephone wires.

Fall was when the geese went honking over, coming from some mysterious place Up North and heading for we didn't know where.

The perch didn't always bite on the minnows we dangled in summer from Sheboygan's North Pier, part of the breakwater system that shelters the Wisconsin city's Lake Michigan harbor. But there were always gulls and sandpipers to watch and to wonder about.

I recall vividly gaining a whole new respect for wild citizens when my dog, Fritz, and I came across an injured horned owl in the woods along the lake shore. Fritz was a pretty tough guy, and he thought maybe he'd take a nip out of that owl. A second later, Fritz had changed his mind and both he and I knew a whole lot more about owls than we had before.

And a wonderful old family story has to do with a squirrel that my mom once coaxed into the house. The animal terrorized her sister, Aunt Nellie, delighting us little rascal kids. It's safe to say that none of us ever forgot that day, especially not Aunt Nellie.

A later story of Marilyn and me and our two kids, Jon and Tony, has to do with the blue jays that nested in bittersweet vines outside a sunporch window in suburban Milwaukee.

One day, it was time for the little ones to leave the nest. Mom and Pop knew it, but the little ones didn't.

We watched in delight as Mom and Pop lured the four little ones from the nest and out along a branch and then called and cajoled them to flutter to the ground 4 feet below.

The terrified little birds teetered desperately on the branch, trying to not disappoint Mom and Pop, who insisted that there was a world outside the nest. One by one, they let go of the branch and joined the Big World as we watched in fascination.

On the other side of that same house, a pair of barn swallows arrived on the same day each spring and built a new nest under the overhang that protected our back door.

We — Marilyn and I, Jon, Tony and Haney the Hound — lived our lives and went back and forth under the swallows living their lives in their parallel universe.

Twenty feet away, under the garage eaves, a colony of bald-faced hornets built a football-sized paper nest one summer. They lived in their universe and we lived in ours. They didn't bother us, and we didn't bother them. But there was not one caterpillar in my vegetable garden that summer.

And so we watch and marvel at those wild citizens whose world we cannot quite enter, and we try to do a little something to make them feel more at home around our homes.

That's what this book is all about.

— Ron Leys

HUMAN NATURE AND THE NATURAL WORLD

One year, a battle raged in the retirement community of Sun City, Arizona. It wasn't about whether to plant grass lawns in the middle of the desert, but about the best time of day to water them. When we are manipulating our landscapes to please ourselves, we can usually accomplish what we want, even when it makes no sense at all.

Occasionally, of course, Mother Nature steps in and thwarts our designs. In Milwaukee, Wisconsin, where the winter temperature sometimes reaches 30 below zero, a neighbor of ours tried to grow roses the way she'd grown them for years at her old home — without any winter protection. Every fall she'd say to me, "Well, we never covered our roses on Long Island." Every spring, she'd buy new rose bushes.

When we turn our thoughts and plans to sharing the landscape with wild birds, animals and insects, we suddenly are forced to think differently, because Nature's other creatures have different priorities, originating in needs, not aesthetics or memories. Four needs: food,

An old apple tree might not seem worth saving, unless it's viewed as a possible apartment house for animals. (Julie Margerum-Leys)

Tangles of plants on the property line of an urban lot can give birds a place to hide. (Sandra Stark)

A non-traditional wren house. The owner hollowed out the corner fenceposts, drilled portals, capped the posts, then watched as wrens moved into three of the four houses. In the garden beneath the fenceposts, his cabbages stayed free of caterpillars until nesting season ended.

water, shelter and space. The space is for what needs to be done — breeding space, brooding space, space to advertise their existence. For that reason, if you want to manipulate the space you own in order to attract wild things, you'll have to think differently. Sometimes as you begin to see new connections between animals and the habitat that sustains them, your new thinking will seem topsy turvy.

You'll find yourself saving dead trees and even creating more of them, valuing brushy thickets and poison ivy, and rating a shrub not by the beauty of its flowers but by the height and tenacity of its fruits. You'll plant berry bushes or fruit trees or cornfields solely so that animals and birds can reap the harvest, and rejoice when you find certain caterpillars chomping on the leaves of some of your flowers.

When you look for new areas to plant trees, you'll seek out the places with the poorest soil. You'll maintain areas not because they are beautiful, but because they are rare. If a lumber buyer offers you a small fortune for those towering oaks in your woodlot, you'll think twice.

It may suddenly occur to you that the "cide" in "insecticide" also occurs in words like "homicide," and that insecticide applies to butterflies as well as those so-called pests. At that point, attracting birds will begin to seem useful as well as amusing. Some of those birds will help you keep the pests in check. Once you've made that decision, it may strike you that improving your soil is also a logical step. Good, rich soil contains earthworms, also known in some circles as "bird food."

You'll begin to have respect for bugs that aren't necessarily beautiful, like the ones that eat or attack the insects that devour the flowers you've planted to feed desirable insects and birds, and the vegetables you've planted to feed yourself. Bees, you'll realize, help those vegetables and flowers by pollinating them. And not killing spiders might even help bring ruby-throated hummingbirds into your neighborhood, for those beautiful little birds are spider-eaters who glue their tiny nests together with spiderwebs.

As your thinking changes, a yard that once seemed perfect, with every dead flower snapped off and discarded, every weed gone, every leaf swept up and toted away, will begin to appear more like a store full of empty shelves. Burning parts of your property will come to seem the best way to improve their value as wildlife habitat. And you'll start to regard weeds differently, too.

If the idea of leaving weeds in your garden makes you uncomfortable, be scientific and refer to them as "forbs," or take up one author's practice of calling them "field species." (37) As for blackberries and other plants that turn themselves into thickets in the blink of an eye, do as a wildlife biology professor once advised his class: don't say "brush," say "flowering shrubs."

This dead tree has proven its value as a cafeteria for chickadees. (Sandra Stark)

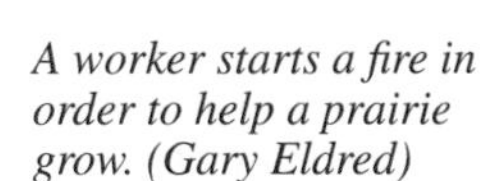

A worker starts a fire in order to help a prairie grow. (Gary Eldred)

Chapter 2

ALL TOGETHER NOW

We all need a place to live. We need food, water, shelter from the elements, protection from those who would harm us and a means to reproduce.

That includes you. It includes me, and it includes the wild citizens of the world.

You and I do a pretty good job of providing those things for ourselves, working at our daily jobs and using our paychecks to buy houses and grub and police protection.

The wild citizens of the world also do a pretty good job, hustling endlessly for the next meal, finding shelter from winter storms, keeping a sharp eye out for predators, and keeping another sharp eye out for mates.

I recall looking out the window of a bush plane flying over the Hudson Bay Barrens of far northern Ontario, watching the endless tundra and pothole country unroll under the plane, seeing no sign of human effects on that land, and thinking, "This is the way the world looked before we screwed it up."

Under those wilderness circumstances, wild animals do just fine. They don't have easy lives, but they live and reproduce and nature sees to it that things stay in a sort of fluctuating balance.

But where humans have changed the landscape in major ways, which includes the cities where many of my friends live and the country where I live, wild animals need some help. That help should go beyond keeping the bird feeders full in winter, as important as that might be. This help should include doing something about habitat, which means providing places where wild citizens can hustle their own grub without handouts from us.

So how are we doing in that department? On balance, pretty well. It's easy to find problems, to worry about logging in national forests in the United States or in rain forests in Brazil, about urban sprawl and highway widening projects.

But we should also remember what we are doing right. For every deep-woods species harmed by logging in national forests, for instance, an open-country species is helped. For every farm that disappears to the developer's bulldozer, another is enrolled in the federal Conservation Reserve Program. That program alone has taken 30 million acres of former cropland — an area about the size of Wisconsin — and turned it into some of the best wildlife habitat in the country.

Most conservation and environmental groups are driven by city people with slightly guilty consciences, and those groups do wonders in pressuring governments to set aside wilderness and natural areas and then to maintain those areas as wildlife habitat.

Urban sprawl creates many problems, but my experience with people who move out from cities to 5- or 10- or 40-acre homesteads in the countryside is that many do a pretty good job of seeing to it that the bobolinks and cardinals and rabbits and deer and butterflies and bumblebees have a place to live and feel safe.

A good friend who runs a small-town cafe proudly showed me photos one morning, photos of wild turkeys strutting through his front yard. Doyle Lewis is as enthusiastic a hunter as he is a Green Bay Packers fan, and I've scrambled across the Colorado Rockies in search of elk with Doyle. But I've never seen him prouder than he was with those photos. Doyle and his wife, Nancy, live in a modest home on the banks of the Wis-

consin River, and they see to it that wild creatures have what they need on the Lewis property.

The truth is that many species of wildlife, including turkeys, are doing much better than they were 100 years ago. They are thriving because of people like Doyle and Nancy Lewis, and because of people like you and like me.

Yes, things could be better. Things could always be better. But we are working to make them better.

As my television hero Red Green says every week, "We're all in this together."

— Ron Leys

WILD THEORY

Before the suggestions begin for attracting wildlife (or for turning off the attractions if you already have too much of a good thing), let's look at some theory to give reasons for the changes you should make to your land. Definitions are in order, too.

HABITAT

A wild creature's habitat is that peculiar combination of food, water, shelter and space that it requires for survival.

Different animals and birds and insects define these four things differently. For instance, for a butterfly, "food" may include nectar from flowering plants or trees; for certain birds, "food" may include that butterfly. But if your piece of land or your land plus your neighbors' properties don't provide the right habitat, you won't be seeing the species you want to see, no matter how hard you try.

Before the end of the 19th century, large expanses of the country were covered by prairie, but when equipment was invented that allowed farmers to plow prairie sod, they quickly realized that this was easier than clearing forest lands, which involved girdling and burning trees and digging up stumps. But the changes the farmers made didn't only destroy the tallgrass prairies. Bobolinks, meadowlarks, dickcissils, upland sandpipers and other prairie birds that had depended on that special habitat disappeared, too. As early as the end of the 19th century, a visitor returning to a former Illinois prairie lamented that he hadn't seen any of the birds of his childhood. The Bell's vireo had disappeared along with the thickets that sheltered it, as had two species of kites. Except in a single remnant prairie of 160 acres where the birds still lived. (75, 56)

Curiously, even the notes a bird sings might have a relationship to the habitat where it sings them. Higher frequency sounds are more likely to be absorbed by forest vegetation, so birds that usually live in forests will sing songs with notes of lower frequencies. The ground also absorbs sounds; a ruffed grouse standing on a low perch means for the sound of his drumming to carry a long distance — and that sound is also of a lower frequency. "Recent studies have shown that the acoustic features of songs used by a species may be closely adapted to habitat characteristics in order to avoid loss of information during singing," write Paul Ehrlich, David Dobkin and Darryl Wheye in *The Birder's Handbook.* (35)

TWO KINDS OF CARRYING CAPACITY

Suppose you decide that you want to encourage grouse to visit your property. How many grouse can you reasonably expect to join you as residents? The answer depends on the biological carrying capacity of your land, the ability of your property to support a given number of grouse. According to Stephen DeStefano, Scott Craven, and John Kubisiak, writing in *A Landowner's Guide to Woodland Wildife Management*, the carrying capacity of a property will always be limited by whichever of the four basic requirements is the most scarce. You could define "wildlife management" and your tasks as a landowner as trying to remove these limits. But eventually, something will put a ceiling on such efforts. For instance, "space" for a male grouse is the territory he will defend from other male grouse. If you don't have enough, you probably won't have grouse as co-owners of your land, no matter how good the food and cover and water are. (32)

By the way, wildlife experts are starting to talk about "social carrying capacity" or "landowner tolerance." This is the population level that a group of animals can reach before the human owners of a piece of property feel overwhelmed. Often, as in the case of deer in residential areas or commercial orchards, the number is lower for social than for biological carrying capacity. In other words, people won't put up with as many deer as the available food, water, cover and space will allow, especially when "food" is being defined as trees and shrubs and plants in which the human owners have invested a lot of money. (79)

The more flexible a bird or animal's habits are, the more likely it is to survive. Robins seem to do better as a landscape changes from farmland to suburb, while animals like salamanders, frogs, turtles and snakes suffer particular hardships as ponds and wetlands where they breed are drained and filled for new subdivisions, or as water becomes polluted by runoff containing lawn chemicals. Yet even if your area is urban or in the process of becoming developed, you might come across a garter snake, for these snakes don't need water and can feed on a variety of things. (62) It's a good bet that creatures you see most and in the most different places are generalists when it comes to the four basic necessities. That is, they are the sort that can get the moisture they need from standing water, dew-laden plants or juicy berries instead of depending on a creek or pond that never runs dry, or relying on your backyard birdbath. (32)

The animals that are fittest can use any of several habitats to survive, while animals that need a single, very specific habitat become endangered if that habitat is beginning to disappear. The ability

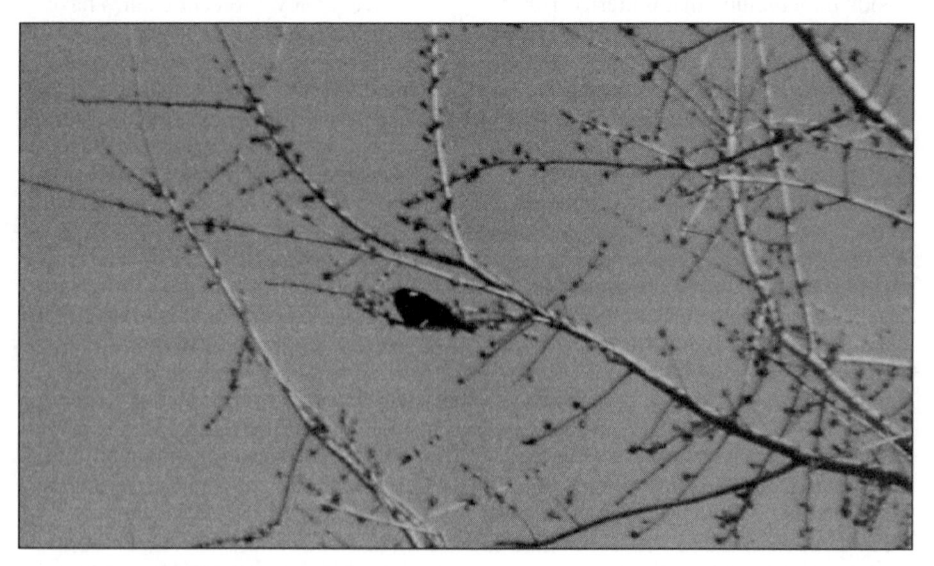

A red-winged blackbird demonstrates the adaptability of his species by making use of a maple tree instead of nearby tall, dry weeds.

to switch from one habitat to another can lead to a rapid increase in the members of a species. Red-winged blackbirds are a good example of this phenomenon (4) — or a bad example, if you happen to own a cherry tree with fruit you were planning to harvest.

EDGE — BENEFITS AND DRAWBACKS

If some species need one kind of habitat and some species need another, it follows that the place on your property where you are likely to find the most different species of wildlife will be where two or more habitats meet. Wildlife specialists have dubbed such areas "edge."

Some species have even evolved to prefer the transition zone or ecotone between two habitats. (107) Song sparrows, brown thrashers and house wrens, among others, have adapted to these areas. (32)

Later in this book, you will read about how to increase edge on your property in order to increase the types and numbers of creatures you can see. But be aware that there are also problems connected with edge, and they stem from the fact that some creatures need large tracts of uninterrupted open land and others need large tracts of forest for protection.

If you look at Table 3-1, you'll find that brownheaded cowbirds are among the birds most often spotted in the listed states. Cowbirds have rapidly made themselves unpopular with birdlovers because of their annoying habit of using other birds' nests for their eggs. Unfortunately, cowbird's fledglings are bigger and will take food from the other young birds in the nest. As a result, songbirds, like the yellow-throated warblers, song and chipping sparrows, scarlet tanagers, red-eyed vireos and eastern phoebes, that prefer large expanses of forest to shield their nests and broods from predators like the cowbird are showing up in smaller and smaller numbers when birds are counted. (57) In all, 121 species are known to have raised cowbirds successfully, which means they have raised their own young less successfully. (62)

One of the culprits responsible for the cowbirds' success is the creation of edge, whether it is done deliberately to increase wildlife diversity or is simply a byproduct of building new houses or parks in the midst of woodland. Fence rows, windbreaks, ditches and travel lanes left over at the edges of farmers' fields do provide a little nesting cover for songbirds, but snakes, foxes, raccoons, opossums, skunks, chipmunks, blue jays and cats use these edges too, and they quickly discover that the songbirds are here; for them, these areas can become prime hunting areas. (57) If a large expanse of forest was available for the songbirds, these predators and the cowbirds wouldn't roam more than 600 yards through the trees in search of eggs or nests. Today, a circle of forest must have at least 250 acres to have an interior safe for songbirds. (62)

In fact, a Wisconsin publication advises that much of the southern part of that state is already so broken up into small woodlots surrounded by large open fields that "if you own a large wooded tract in southern Wisconsin, you should consider preserving this unique community," instead of increasing the amount of edge by creating openings. (32)

SUCCESSION

The most dramatic examples of edge are places where forest meets field or water meets any dry land habitat or urban lawn meets a row of trees. But the edge effect also occurs where habitats in any two stages of succession come together. (89)

Succession is the name wildlife experts put to Nature's habit of never standing still.

If a farmer plows a field in an area where woods predominate, then abandons his land, it will not stay bare. In the first year, annual and perennial weeds move in. Within two or three years, the perennials will crowd out the annuals, but eventually seeds of shrubs and trees will either be blown in by the wind or carried in by birds, who will eat the good parts of the fruit they find and inadvertently plant the leftovers. These will be seedlings of trees that need direct sunlight to grow — hawthorn, aspen, cherry, birch, white pine and others. After about 10 years, the brush will have grown into sapling-size, then pole-size trees whose leaves and twigs deer will not be able to reach and browse. (43)

In 20 or 30 years, these seedlings will have grown into a woodlot, but because they shade the ground, their own seedlings will not thrive. Instead, beneath these mature trees, the seeds of shade-tolerant brush and trees will be taking hold. When those trees grow to dominate the woods, they will be able to self-seed and the woods will be said to be a climax forest. (32)

If your forest is in an early stage of succession, deer mice, ground squirrels and white-crowned sparrows will be among the creatures living on your land. (89) Juncos, Carolina wrens, catbirds, cardinals and towhees thrive in the young forest where they can find food and cover on the ground and in the underbrush. Later, pine martens, red-backed voles, goshawks, hairy woodpeckers, ruby-crowned kinglets and hermit thrushes will be among your neighbors. (89)

In an area where grassland prevails, succession will look a little different. Here, annuals and perennials are succeeded by short-lived perennial grasses, which eventually will be succeeded by sod-forming grasses. (89) Ponds and lakes, which also feel the effects of succession, provide different definitions of "water" as the years go by.

If you are seeing less of a favorite species on your half acre, consider whether that species was relying on some plant or tree or clearing that has changed its size or even disappeared. You might have to examine a neighbor's property to solve the mystery. The larger the area you own, the more species you will be able to attract, because bigger areas are more likely to contain a mixture of stages of succession and larger patches of each stage. (30)

But whether you own half an acre or 40 acres, succession will constantly change the way your land looks, and the way your land looks will determine the animals that come to visit. Unless, of course, you intervene to maintain a desired stage of succession.

PREFERRED STAGES OF SUCCESSION

Some stages of succession are better than others for supporting wildlife. You'll see the most species from the time a piece of land is bare through the period in which trees are still small enough to let a lot of sunlight through. That sunlight is important because it encourages the growth of shrubs and other ground cover that many animals thrive on. (89)

You'll see the least wildlife in the next stage of succession, as the trees grow tall enough to evade the reach of browsing animals, yet shade the forest floor so heavily that little or no understory can grow. (43)

When the shade-tolerant trees appear, you'll start to see a little more wildlife. But in a mature or overmature climax forest, you'll have the second best chance of seeing a variety of species, since some trees will begin to die, leaving clearings where succession can start all over again and wildlife again has an ample supply of food. (89)

HORIZONTAL AND VERTICAL DIVERSITY

Clearly, diversity of species depends on a diversity of plants and shrubs and trees. Another way of looking at diversity is by considering horizontal and vertical diversity.

Horizontal diversity means that within the land you own, you have different patches, each one offering a different type of food, water, space and cover, so that an animal or bird or insect doesn't have to go far in order to have all of its needs met. Even on half an acre, an open space with wildflowers plus some shrubs and trees for cover can provide horizontal diversity. (4)

Vertical diversity starts with the idea that different creatures like to do things at different heights, with their differing temperatures, humidity levels and menus of insects and other available food. (4, 35)

A study by Robert MacArthur showed that five different species of warblers — Cape May, yellow-rumped, black-throated green, Blackburnian and bay-breasted — that appeared to be eating from the same spruce, fir and pine trees were actually feeding in different areas of those trees. For instance, Cape May warblers mostly dined on the new needles and buds on the outside of the tops of the trees, while the bay-breasted

The dramatic difference in height between plants and trees leads to diversity in the wildlife they attract. (Sandra Stark)

When the heights of flowers vary, the wildlife they attract will vary too. (Sandra Stark)

warblers concentrated on dead needles toward the center of the middle of the trees. (35)

In a forest, insects and earthworms in the duff attract moles, shrews and nesting birds. Ferns, mosses, grasses and wildlowers provide low ground cover for hermit thrushes, insects, box and wood turtles, mice, snakes and toads.

A little higher up, in the shrub layer, no more than 15 to 20 feet above the ground, viburnums, mountain laurel, winterberry, huckleberry and blueberry draw in such songbirds as rose-breasted grosbeaks and cardinals that build nests in the shrubby thickets. (67)

In the layer of small trees typified by flowering dogwood and hornbeam, other songbirds nest. In the canopy formed by the tallest of the trees, insects on the foliage attract warblers, vireos, flycatchers, scarlet tanagers and other insect eaters. (67) And on tree trunks of all sizes, nuthatches, brown creepers and woodpeckers patrol crevices in the bark. (43)

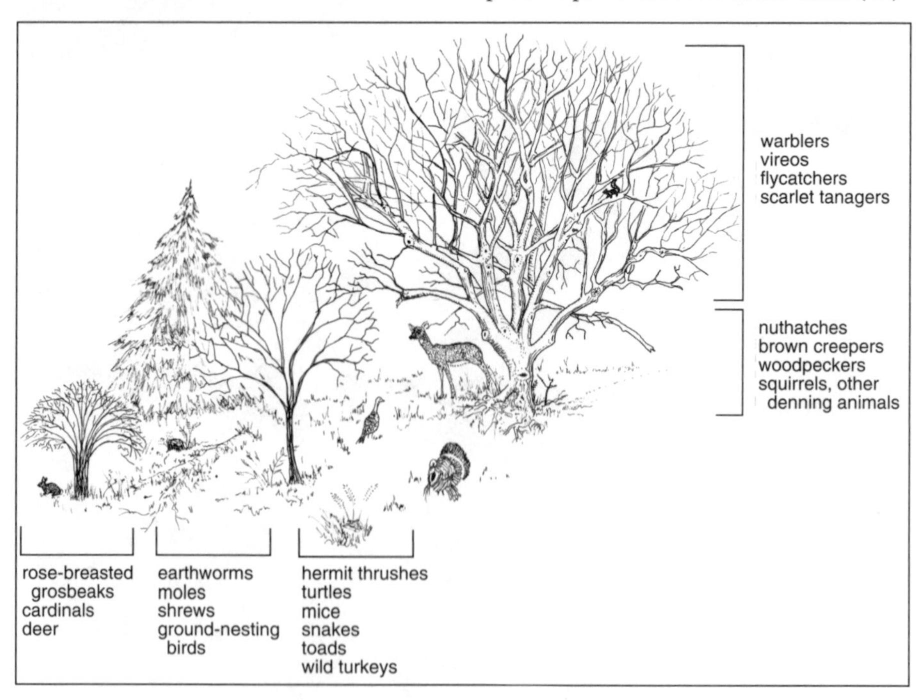

Different creatures seek out the varying habitats that occur at different heights. Although birds that migrate in winter would make use of the trees when they are in leaf, hardwoods have been left bare to show their structure, along with holes in the large wolf tree at right. (Patricia Gilbert)

Clearly, the more vertical and horizontal diversity you have on your property, the more wildlife species you'll be able to invite. Besides, growing only one or two species of trees or plants increases your chances of being wiped out by disease or, at the very least, set back by a crop failure when blossoms are frozen out or acorns fail to form.

Even in a very small urban space, it is possible to supply diversity. For instance, a Vermont author reminisces about a house he lived in with a back yard 30 feet square. However, a tall tree filled most of that yard and a brush travel lane connected the back yards on the block, so there was wildlife to enjoy. (87) Because such a large percentage of the residents of New York State live in extremely crowded urban areas, the Cornell University Cooperative Extension Service's *Wildlife in Today's Landscapes* has suggestions for rooftop gardening to attract wildlife. (62)

INCREASE WILDLIFE BY IMPROVING HABITAT

When a species becomes endangered, some people automatically blame human hunters or other predators for the disappearance of the wildlife; however, there is strong evidence that the quality of habitat, not the presence of predators, determines whether an animal or bird is successful. When Indians hunted ivory-billed woodpeckers and used their bills for decorations, the birds still thrived. In fact, they did not begin to disappear until the timber industry began the wholesale removal of the undisturbed, old-growth swamp forests that were home to these birds. Unfortunately, the ivory-billed woodpecker has a broad definition of "space": Each breeding pair needs three square miles of undisturbed forest. (35)

If you improve the quality of the land you own by providing the four things wild creatures need to survive in diverse environments, and you take into consideration succession plus the effects of your efforts on all of the wildlife that is or might be on your property, you stand a good chance of succeeding in your work.

Chapter 3

STOPPING, LOOKING, LISTENING

It was our first trip to southern California, and Marilyn and I came armed with binoculars and a bird book. We spent many hours on Pismo Beach and the surrounding area, arguing over whether that gull overhead had yellow legs or orange legs, a solid-colored or striped bill, the leg and bill colors giving clues to which species the gull belonged.

Our son Jon said of this, "Julie [his wife] and I never pay any attention to whether a bird has yellow legs or orange legs, or a striped or solid bill. We just say, 'Isn't that a pretty bird?'"

"Oh," I said, not having a more articulate answer.

Until the next day.

We were stuck in traffic, when Jon said, "Hey, look at the beautiful '69 Mustang."

So I innocently asked, "How do you know it's a '69 and not a '68?"

"Well," he said, in that time-honored tone that young men take when they are addressing a dull-witted parent, "the '68 has square taillights, while the '69 has round taillights." Or maybe it was the other way around.

Anyway, I said, still Mr. Innocence, "Your mom and I never pay much attention to whether a car has square taillights or round taillights. We just say, 'Isn't that a pretty car?'"

There was a long silence. Then Jon said, "You got me."

"Yes, I did," I replied.

Such moments of triumph over a smart-aleck kid are rare, and all the sweeter for it, and the incident has become part of our collection of family stories.

But it also illustrates how one begins noticing details, looking for them and remembering them, as one's interest in a subject deepens.

Jon was more interested, at that point in his life, in Ford Mustangs, and I was more interested in Pacific Ocean gulls.

During the 1950s, I was something of a hotrodder, at one time even owning a red Ford Deuce roadster, and to this day when I spot a Ford of 1928 to 1940 vintage I can tell you exactly what year it was made and how I know.

But these days of my life, I find more interest in how evolution or God, take your choice, shaped the sandhill crane's silhouette a little differently than that of the great blue heron than in how Henry Ford changed hubcap sizes from the 1929 to 1930 Model A.

"What difference does it make?" you might well ask.

To some, being able to identify a bird by species is their only goal. Some birders become what the British know as "tickers," interested only in ticking off a word on a list.

On a birding excursion in the area of Aransas Pass on the coast of Texas, I heard a woman say to another, "What kind of bird is that?" "A wood ibis," was the reply. "Oh," the first woman said in disappointment, "I got one of those this morning."

It's safe to say that she will never know much about the wood ibis.

On the other hand, if I had not for some reason memorized the different shape and stance of the sandhill crane, and associated that shape and stance with the name of the bird, I would never had noticed that a pair of cranes had moved into pastures along Richland Creek, just north of the Wisconsin River near our farm. I would have assumed they were just some more blue herons.

I would not have thrilled to see this magnificent bird extending its range into my neighborhood, the same bird that Aldo Leopold, who taught us how to love this Earth, predicted during the 1930s would very soon disappear.

Just so, if I were not able to tell a turkey from a crow at 100 yards, I would have missed the firsthand knowledge of how the sport of hunting and the fees collected from hunters brought an interesting bird back to Wisconsin after a 100-year absence.

But those are some obvious examples of birds with obvious differences. It's like being able to pick out the sound of the clarinet from the sound of the saxophone in a jazz band. But if you listen to jazz long enough, and if your interest deepens, you'll be able to pick the alto sax from the tenor sax, the trumpet from the cornet. Just so, if you pay attention to the birds, there will come a day when you can tell a ruby-crowned kinglet from a golden-crowned kinglet.

Again, is this important? It is, let me assure you, to the kinglets of the world. They, after all, consider themselves very different animals and they would be insulted beyond belief if they even suspected that some people did not believe it important.

But going beyond the opinion of kinglets, it's a way for us to begin to know something about wildlife, to organize our thinking, to understand that birds that might look somewhat alike are really very different, with different preferences in food, in shelter, in general areas where they feel at home, in behavior patterns.

To know that a piece of wood is oak or pine is just the beginning of knowing anything about building a house or a bookcase, but would you hire a carpenter or a cabinetmaker who couldn't tell oak from pine?

So get out that bird identification guide, that pair of binoculars you got for your birthday, go sit on a stump or a stool, and pay attention. You will learn something. I guarantee it.

It's not really that hard. Until you get to the kinglets.

— Ron Leys

LEARNING TO SEE

Once you've decided to go ahead and do something to attract wildlife to your land, waiting does not seem logical. But in this case, waiting is the best first thing you can do, as long as you watch while you wait.

What follows are suggestions of things to look for on your land and that of your neighbors. These are places where you're most likely to find specific animals and birds, and ways to organize your observations. Not all of the suggestions will apply to the land you own, but some of them might. And some might be things you've never given much thought to — keeping your eyes open for owl pellets, for instance. If you can keep your enthusiasm in check while studying the local vegetation and wildlife in all their seasons, you are more likely to know where to begin and less likely to evict someone you might want to remain.

WRITTEN RECORDS

A good way to stay organized as you wait and walk around and watch is to keep a diary of what you see and what the weather conditions are when you see it.

Whenever you walk, notice which way the wind is blowing. You'll need to know which sides of your house are most exposed to the cold of winter if you decide to plant a shelterbelt — an arrangement of trees and shrubs that will protect birds and animals in addition to cutting heating costs for you. You'll need to identify any leeward walls that might shelter bird feeders in winter, and south-facing walls that could give wildlife-enticing plants a head start in spring.

If one of your aims is to attract butterflies, you'll need to know something about the prevailing summer winds in order to locate buffers where they will do butterflies the most good. (37)

A checklist is an alternate way of keeping your records. You can keep track of broad categories like "game animals," "songbirds," "hawks and owls," or whatever you are interested in, then the time of year you happened across a given creature and where you saw it. (43)

SURVEYING THE NEIGHBORHOOD

If yours is an urban or suburban lot, broaden your walks to include the area beyond your property and your immediate neighbors. Are there any undeveloped parcels within a half mile or mile? Any public parks, estates, cemeteries, golf courses that might be sheltering birds?

Get a feel for the dominant habitat in your neighborhood. Is most of the area open? If there are a lot of trees, are most of them mature or young? This way, you'll be able to plan additions to your yard that are in harmony with the way things are. For instance, if your small urban lot is in an area shaded by many mature trees, plant young trees of the same species and eventually you'll attract the kinds of birds and animals that are already making use of that habitat. (53)

If your property is small, it's obvious that what your neighbors are doing will affect the wildlife that visits you. But even on a piece of land 40 acres or larger, you need to discover which of your neighbors is providing which of the four requirements — food, water, cover, space — so that you can concentrate on providing what is missing. If you decide to harvest an acre of mature trees in order to encourage brushy cover for grouse, and at the same time the farmer to your north clears ten acres of brush to make a new pasture, the result may be a net loss of grouse, plus a lot of wasted work on your part. (43)

Wild turkeys need both woods and open fields to make a living, and they routinely range over an average of 1,500 acres. Obviously, then, if your goal is to attract turkeys to your country acreage, you need to study the entire area around you. Aerial photos obtained from a county Farm Services Agency (FSA) office will help, as long as you remember that a photo only tells you what the area looked like at the moment the photo was taken. You'll also need to know what the area looks like during other times of the year. (70)

If you own property in an area that has remained relatively undeveloped, assign yourself the job of discovering what happens naturally on your land and the land that adjoins it.

Try to find natural openings, spaces where less than 10 percent of the open area has trees and less than 30 percent is covered by brush. Depending on what the land was used for in the past, these openings might have been the result of an established homestead that was subsequently abandoned, or loggers might have created an opening to stack logs where truckers could load them easily. Or, it might simply be a low-lying frost pocket, where plants have a hard time establishing themselves. An aerial photo can be used to identify openings as small as three acres. (89)

If you find such an opening, identify which grasses, grains and/or legumes are covering the surface that isn't occupied by trees or brush. At home, where you want to create a similar opening in a similar spot, that type of herbaceous cover may well be the easiest to grow. Look also at the forbs — wide-leaved plants — and shrubs. They'll give you a clue about natural succession in the area, and the kind of competition to be encountered and, possibly, eliminated for any young trees you plant. (89)

Perennial forbs die back in winter, but grow the next season from the roots that remain, while annuals depend on spreading lots of seed that will sprout when conditions are right. Therefore, annuals are a lot more useful to seed-eating birds. As you examine the natural openings in your area, notice whether any annual forbs remain, or whether

Milkweed is the annual forb of choice for monarch butterflies and their caterpillars.

Take note of the weeds on your property. Goldenrod attracts many kinds of butterflies, including clouded sulphurs like this one.

succession has advanced to the point where the perennials have entirely crowded out the annuals, a process that will usually take only two to three years unless something has disturbed the soil.

Where an opening has been caused by wind damage, notice how big that opening is. Later, if you decide to clearcut part of your woods to stimulate the growth of herbaceous cover, you should be aiming for an opening no bigger than what a natural disturbance would supply. (89)

If your property is a small urban or suburban piece of land, your search for native grasses and forbs should take you to any undisturbed bit of land you can find. Maybe there's a hillside too steep or a ditch too moist to mow, or one of your neighbors has decided to let part of his lawn revert to whatever grows wild. Maybe a vacant lot or an out-of-the-way corner of a park can supply you with some clues. Some time ago, an article in *Smithsonian* magazine chronicled the detective work of an organization dedicated to the preservation of antique varieties of roses; the members haunt old cemeteries, the most fertile grounds for what they seek.

On any size property, look for wild grape vines, a good source of food for birds. If the vines are using neighboring trees for support, and those trees are in danger of being shaded out by taller trees surrounding them, you might decide to help the natural grape arbor by cutting the shady characters. (89)

MAPPING YOUR DISCOVERIES

In addition to keeping a diary or some other written record of the things you see, you should also sketch a map of your property. If you have a property survey, or if you can get one from your municipal government, you'll know immediately the dimensions of your urban or suburban lot and the location of the buildings on it. (111) For larger properties, aerial photographs available from the Agricultural Stabilization and Conservation Service (ASCS), perhaps available from your county agriculture agent, or topographical maps from the U. S. Geological Survey will help give you an overview of your land. (43)

Your personal map of your land should show permanent features that you can expect to see year after year — not only buildings, but trees, shrubs and perennials that are already on the property. If there is an area where the soil has been improved for a garden, indicate that on your map. Don't forget to include notations of water sources, spigots or sprinkling systems. Add notes on pleasant views, plus anything you might want to screen so you don't have to see it every day. Put in anything that might act as a windbreak to shelter wildlife, like a hedge or fence. Examine and put on your map anything in your neighbor's yard that might shade yours, plus features like trees whose benefits you might want to continue across the property line with plants in your yard. (111) And add an arrow to indicate where the prevailing wind comes from.

MAPPING A LARGER PROPERTY

For the map of a larger property, use buildings, driveways, fence lines, woods, trails, streams and ponds as landmarks to start from. Then sketch in other features of use to wildlife, like pastures, crop fields, orchards, tree plantations, mixed forests, shrub hedges, fence rows of mixed trees and shrubs and/or utility right of ways that might also be serving as travel corridors, as well as nurseries for plants that feed wildlife. (43)

Record information about any stands of trees on your land. Stands are blocks of trees that have something in common, like the stage of succession they are in. If you can't guess the age of the trees in a stand, include the DBH (diameter of the tree at breast height, or 4-1/2 feet from the ground) or the estimated height. As already mentioned, different birds and animals live in different ages of forest. If your forest or your neighbor's woodlot is the wrong age for the animal you want to attract, you won't be seeing that animal, but there are things you can do to alter the situation, and you'll read about them later in this book.

If there are single trees within the woodlot that are taller than the rest, locate them on your map. These are bonanzas for birds of prey, who like to perch above the forest canopy in order to scout the territory. (43)

Record the slopes and exposures of your land. Some trees and plants simply will not thrive on a north-facing slope (32) and a steep slope that shows signs of erosion might be a priority project for you.

If your neighbor is growing crops, make a record of what they are and when they are harvested. Standing crops feed wildlife as well as humans. If your neighbor is going to foot the food bill, you might as well take advantage of it too. (32)

MAPPING SPACES FOR PRAIRIES

If you own a small property, map any slope too steep to mow in comfort. Some day, this might make a good place to put a miniature prairie. Once it's established, it will cut down on your yard work and act as a living privacy fence in summer, since many prairie plants are quite tall. If you do, though, you might decide to follow the advice of Craig Tufts and Peter Loewer, authors of *Gardening for Wildlife*: once you've installed your prairie, put up a sign that says something like "American Prairie Garden" to explain to your neighbors that you're not just encouraging weeds. (111)

If you own a somewhat bigger property and your aim is to re-create a native prairie, add to your map any areas that are flat or slope gently the way the original grasslands did. The areas you map should have some variation in elevation so that a number of different kinds of flowers can be seen clearly. The land should receive sunshine for at least three-quarters of the day, because prairie plants need a lot of sun. (75)

One of the features that settlers remarked on when they first caught sight of the native prairies was the beauty of the tall prairie grasses moving gently in the wind; your prairie site should be getting enough wind to reproduce this effect.

If your property is big enough so that you can carry out a prescribed burn on your prairie, be sure to map buildings, fences, utility poles and other features that don't benefit from fires. (75) And don't set any fires unless you know exactly what you are doing, have enough help and equipment and have alerted the proper authorities.

HIDING PLACES

Prey needs to hide from predators. This holds true whether you're living in a city or in the country. Look for travel corridors — long, narrow passages of brushy cover — where birds and animals move safely from one type of habitat to another. If there's an area where the previous owner has simply dumped rocks and tree limbs, regard it as a treasure rather than a chore to be faced in the future. Unless, of course, it is located very close to a place where you or your neighbor maintains a vegetable garden. Add these sheltering spots to your map.

If you find brushy thickets, stand quietly and watch for a while. You might be surprised at how much wildlife is already on your property. If you've made "cleaning up" the area a priority, you might want to reconsider, especially if your land holds the only brushy escape corridor in your neighborhood.

If your land and your neighbors' are big, open fields or, for that matter, well-mowed lawns with no continuous areas where animals can get around safely, the area is probably short of wildlife as well.

On a larger property, if you want to attract game birds, you'll definitely need to determine whether there's adequate cover in the area. The Minnesota Department of Natural Resources advises that waterfowl need four acres of upland cover for each acre of wetland, while pheasants need at least 30 acres of undisturbed nesting cover per square mile. (83)

Slash happens on small and big parcels of land, for slash is the accumulation of branches that are downed by wind and weather, the limbs and twigs that neat homeowners may feel compelled to remove. But slash can be valuable as cover for game animals and small mammals. It can provide cover for small birds and attract insects that will feed those birds and the predators that dine on them. (89)

MAPPING TREES

What kinds of trees do you have on your property and how are they arranged? On a small property, you ought to have one evergreen or more. Red squirrels, if you have them, will eat the seeds of various evergreens. Mourning doves will make their nests in evergreens, though they'll sometimes greet a spring morning with their characteristic cooing a bit earlier than you might prefer. Songbirds use evergreens to shelter

In winter, trees like these will shelter a multitude of birds.

It may look unkempt, but a staghorn sumac thicket offers cover plus food for wildlife.

from hawks. And deer browse conifer needles in winter.

On a larger woodlot, in each acre of deciduous woodland there ought to be several clumps of about six shade-tolerant evergreens, like spruce, spaced 6 by 6 feet apart, to provide winter cover for birds. (43)

To be of value to birds, an evergreen's lower boughs ought to be close enough to the ground to provide shelter. A pine plantation where Nature has trimmed all of the lower branches may seem a thing of beauty to the human eye, but it doesn't offer much in the way of winter protection.

Look for and map any mast-producing trees, producers of acorns, nuts and seeds. Count your search a success if you find oaks, hickories, beeches, walnuts, butternuts, cherries, ashes or evergreens that still have their lower branches. Look, too, for birch, hazel, alder and aspen, whose male catkins will be eaten by some wildlife. (89)

Acorns are by far the most popular food for a large variety of wild animals. Big trees will produce more mast than smaller trees, though oaks bear the most seed at middle age. Look, too, at what is growing under the trees that you're planning to keep because undergrowth will also feed the wildlife in your area. Generally, it's important to provide browse beneath mature trees. (89)

But if the understory contains an invasive shrub like bush honeysuckle, one of your first priorities might be to rid your woods of it. Bush honeysuckle is easy to recognize; it will be the first green in your woods, if it's there. That's

Bush honeysuckle (Bureau of Endangered Resources, Wis. DNR)

one of the secrets of its success: it gets the jump on everything else in spring, and dies out after almost everything else in fall. It's a problem to those mast-producers you want because its shade will kill any new seedlings that try to grow.

As summer becomes fall and then winter, keep track of which foods are still available for birds and animals that winter over in your area. If the neighborhood larder is bare by January, you might decide to start your planting program with bushes and plants that retain their seeds and berries longer into the winter.

If your goal for your land is to attract a specific animal or bird, then you might want to narrow your search by checking for those trees, shrubs and herbaceous plants that they like.

If ruffed grouse are the birds you seek, look for aspen, berry-producing shrubs, prickly ash, hazel, sumac and/or locust thickets that provide the vertical cover grouse like, small forest openings, grape tangles and/or clumps of young conifers. Take notes, too, on the ages of the stands of aspen. Grouse need three different ages of aspen within six to 10 acres to provide year-round food, cover and space for a male to announce his territory. An oak and hickory forest will provide food and cover for grouse, too, if it has an understory that is a mixture of low evergreens, tall shrubs and herbaceous plants. Also look for travel lanes. If there are large expanses of fields with no way for grouse to move safely, there might be no visitors. (32)

If your goal is to encourage deer, be sure to look at what borders your property. If your neighbors have mature timber with little brush in the understory, you might decide to harvest some of the timber on your land to encourage brush and small trees that the deer can browse. If your timber is not mature, you might want to create an herbaceous opening to provide food. If your neighborhood is nothing but open fields, you might want to start an evergreen plantation for winter cover. (43)

If there are already a lot of shrubby thickets, but you find few herbaceous openings, you'll waste a lot of money if you buy shrubs for your property. On the other hand, food plots, corn or other grains left standing above the snow in winter, must be located close to good wildlife escape or resting cover like dense woods or cattail marshes, or you're not likely to see the ruffed grouse, songbirds, deer, squirrels, quail and turkey you're trying to benefit. If you haven't found good places for wildlife to hide as they travel, planting a food patch should not be your first priority. (32)

SAVING DEAD AND DYING TREES

As you walk and watch, pay particular attention to features that standard logic tells you to get rid of, and be sure to add these features to your map.

Snags are at the top of the list of benefits that don't seem beneficial. Snags are dead or dying trees 20 feet or taller that are still standing. The U. S. Forest Service defines a snag worth saving as being more than 16 inches at breast height with more than 40 percent of its bark still intact. Most people think of firewood when they find a dead tree, but

A snag offers opportunities for food and housing.

Over-enthusiastic pileated woodpeckers have finally managed to halve a snag.

if you want to attract wildlife, a snag left uncut offers many benefits — unless it's dangerous because it's located too close to a path that's heavily used by family, friends, and other humans.

Snags are apartment houses for woodpeckers that excavate their own spaces and for other animals and birds that move in when the woodpeckers move out. Woodpeckers are worth pampering because they will search for insects in your healthy trees as well as in dead and dying snags.

Be especially protective of a tall snag, because the bigger it is, the longer it will take to deteriorate. A tree that is dying because its top was broken off will remain standing longer than a snag that still has all its branches, because the broken tree will be more resistant to high winds.

If you own a woodlot with a lot of young trees, you might find remnant snags and trees that were left behind after fire, insects or disease claimed the original forest. (89)

If you own a small parcel of land, look for snags on your neighbors' properties as well as your own. Saving a dead tree is viewed by some people as a very odd idea; a neighbor who has a snag might not be enthusiastic about keeping it, (29) but you might be able to change his mind.

A tree that is merely dying can also be an asset. A tree that's oozing sap probably has fungus-caused heart rot. Woodpeckers know that, too, and will search out a tree like this one because the disease softens the tree and makes it easier for the bird to excavate itself a home. (89) The sweet ooze will also attract butterflies by day and moths by night. (111)

Wolf trees in a woodlot are also worth mapping and cherishing, especially if they are oaks or other trees that provide food for wildlife. Wolf trees have no commercial value as lumber, no matter how big they are, because instead of one straight trunk, they have many limbs. They often start growing in the open, then other trees grow up around them, but the wolf trees stick out above the rest, preying on their neighbors by shading them out. (89) Hence, the name. Their abundant limbs offer perches and cover, while their ability to produce food for wildlife is usually unequaled by smaller trees in the area.

Worth noting, though not necessarily worth mapping, are areas where leaf litter accumulates. Standard logic says that piles of dead leaves are suitable only to rake and bag. However, dead leaves are happy hunting grounds for birds like towhees, thrushes and fox sparrows that will root through leaf litter looking for insects and fallen seeds and berries. Larvae, including some butterflies, winter over in leaf litter, too.

Taking a Wildlife Census

While you are studying the lay of the land, you should also take a census of the birds, reptiles and mammals in your area. To do this, you need to know where and when to look.

As you explore, you'll probably start collecting a reference library filled with various field guides and other books. A word of caution, though: try to get the most recent books you can find. Otherwise, you might be handed outdated advice. For instance, much of the information in *American Wildlife & Plants — A Guide to Wildlife Food Habits*, though published in 1951, is repeated in books published far more recently. It seems that no one has done better research on the stomach contents of birds and animals. However, since 1951, house finches have taken up permanent residence well east of the area indicated in the book, and multiflora rose, extolled in the book, has become a public nuisance.

One way of discovering birds is to know ahead of time which ones you're most likely to see. To do this, first check Table 3-1 to see which birds are most likely to turn up in your state. That's not the end, though, because if you don't have the required habitat in your immediate area, you're probably not going to have the bird either. So check Table 3-2, as well.

Once you know which birds are likely in to be in your area and you have some idea of the kinds of places they frequent, look at a bird guide for a two-dimensional picture to store in your memory, plus information on the bird's typical song and, sometimes, its flight pattern. Knowing who to watch for should make it easier to spot birds as you walk and watch.

SPOTTING SPECIFIC BIRDS

The results of a pair of studies done in Amherst, Mass., seem to offer a good prediction of the categories of birds you might come across if your property is in or near a city.

In the first study, Richard DeGraaf, a wildlife biologist with the United States Forest Service, compared urban and suburban neighborhoods. There were three times as many trees in the suburb, offering food and nesting sites, and, perhaps not surprisingly, almost three times as many bird species. In the suburb, 50 different kinds of birds were breeding, while in the city, 19 species were found.

Then DeGraaf compared three different suburbs. One was a subdivision of similar, boxy houses that had been built in an open field about 20 years earlier. One was a subdivision that had been built at about the same time, but in an existing woodland. The third was an area of homes that had been built at the turn of the century. That neighborhood featured mature street trees, oaks, elms and maples, shading ornate Victorian houses.

Each of the three neighborhoods had the same number of species of birds, but different species had settled in different neighborhoods.

The neighborhood of tract houses in the open field had the fewest insect-eating birds, for clipped lawns provide few opportunities for insects to thrive. The fewest cavity-nesters lived here, since there were few trees big enough and old enough to provide places to build nests. Here, most of the species were ground-nesting, seed-eating birds.

The largest number of insect-eaters and cavity nesters had gravitated to the neighborhood built in the woodland.

As for the old neighborhood, birds that typically eat either grain or grain plus insects were found in the largest numbers here. The cavity nesters like house sparrows and European starlings greatly preferred this area to the neighborhood of boxy tract houses, not, in this case, because there were more trees, but because the ornate trim on the houses themselves became the birds' preferred nesting sites. (62)

Whether you live in or near the city or in the country, if you locate woodpeckers on your property, you've found fine neighbors who will patrol your trees for insects, free of charge. Watch for them as they crawl up tree trunks searching for dinner, or follow your ears when you hear a continuous pounding in your woods or in the direction of a nearby telephone pole; either you're hearing a bird who is hunting, or a male woodpecker placing his version of a mate-wanted ad.

The species that appear in your neighborhood might be pairing up in order to engage in what wildlife biologists call "commensal feeding." This means that one species will follow ("attend") a second in order to eat the first species' leftovers, though the first species has no intention of feeding the second. If you find that a sapsucker has been drilling holes in the bark of one of your trees, you might also find a warbler or kinglet drinking the sap oozing from the "well." Bluebirds and nuthatches will also follow woodpeckers to clean up any overlooked insects.

When a pileated woodpecker visits, it will use its heavy equipment to peel the bark from trees. On such patches, you might find a much smaller hairy woodpecker taking advantage of the insects the pileated woodpecker has uncovered.

If your neighborhood includes a pond or stream, and you find a flock of canvasback, mallard, pintail or redhead ducks or tundra swans, you'll often see coots as well. That's because the coots are taking advantage of the prey that the other birds have stirred up, then ignored. (35)

Birds also benefit when humans initiate the commensal, inadvertent, feeding. City birds learn quickly that wherever people gather, there's likely to be discarded food. If you're looking for birds in a city, look around ballparks or other places where the remains of hot dogs and hamburgers might be. Where winters are cold and snowy, look in areas with underground heating ducts that have plants growing on top, or under streetlights, which extend the birds' foraging hours. (35)

Table 3-2 also shows you what various birds eat. If you come across an obvious food source as you walk and watch, you might also spot the bird that dines there, if you take the time to wait quietly.

MORE SPECIES, MORE PLACES TO LOOK

If you find a tree leaning over water, look on the underside of an outstretched limb for a hole that indicates a woodpecker's nest. As a bonus, you might find that a wood duck has moved into an abandoned pileated woodpecker's nest located over water. Holes like these protect birds from rain and possibly from predators approaching them over water. (29)

In winter, especially if there are a lot of birches in your area, watch for redpolls. You might see them as they knock the seeds off birch catkins, collect the seeds from the ground, then, with the seeds safely stored in special structures in their necks, fly to shelter and protection from predators, where they can, finally, shell and eat the seeds. On cold, windy winter days, both redpolls and crossbills will shelter themselves in dense evergreens, remaining still to conserve energy and fluffed out to increase the insulation value of their feathers. (35)

If you wonder whether wild turkeys are making a home near yours, watch for scratches in the forest duff where turkeys have been looking for acorns or other food items. (68) In winter, drive around the countryside and look for flocks of turkeys in farmers' fields, perhaps following a manure spreader to scavenge leftover corn kernels or even, sometimes, at outdoor feed troughs.

What time of year and day can you expect to see certain species? As you see on Table 3-2, many of the birds that spend some time in the North also migrate. Birds that migrate a long distance usually fly at night, landing, if they do, at about sunrise to rest and search for food. Birds that migrate short distances are usually daytime travelers, and usually fly in the morning. Birds that catch insects on the wing, like swallows or swifts, eat while they travel. (35) As birds begin to migrate, watching for the various species at the times of day when you know they'll be around can be a breathtaking experience, or one straight out of Alfred Hitchcock, if your mind is built that way.

If a migrant finds your property one spring, what is the likelihood that it will be there the next year? According to the authors of *The Birder's Handbook*, a species that nests in a relatively stable habitat is more likely to return to the same place year after year. Thus, you're more likely to find barn swallows in the same locale, barn, garage or eaves, year after year than you are to find bank swallows in the highly erodible river banks that they inhabit. (35)

LISTENING

Your ears are useful for finding other creatures besides woodpeckers. Night is the time to listen for frogs and owls. (111)

You might hear spring peepers trilling late in afternoons and on through evenings in late winter and into spring. On a warm, humid late spring or early summer evening, you might hear the high-pitched, loud call of the gray tree frog, unvarying in tone. All this frog needs is a small, temporary body of water where it can lay its eggs, plus trees where, as its name suggests, it lives. That its home is not in a permanent lake or pond is protection for the gray tree frog and other creatures like it, for it doesn't have to worry about predator bugs and fish that develop in water that lasts. Don't let the name fool you, the gray tree frog, though only 1-1/4 to 2-1/4 inches long, can change color, from charcoal to cement gray, apple or blue-green, very pale yellow or brown, depending on the surface it's on. Or it might be a mixture of these colors when you see it. So probably the sound it makes, rather than its appearance, will let you know what you're looking at. (34)

In a wooded area, a drumroll that sounds like a chainsaw starting up is probably a male ruffed grouse advertising for a mate or three. He has spread his tail and pressed it against a log or some other elevated platform. What you hear is the sound of his wings as he makes a series of strong wing strokes. He'll repeat the series every three or four minutes, and the thumping can be heard for a quarter of a mile or more. You're especially likely to hear this noise just before dawn. (131)

If you actually see a great horned owl, you're luckier than most, though

Barn swallows will return to nest in man-made structures, whether they are welcome or not. All it takes sometimes is an open garage door.

you might well hear one. Its hoot is easy to confuse with a mourning dove's coo, but it's harder sounding and generally occurs at a different time of day. By the way, if you smell the odor of a skunk briefly in the evening, but the smell doesn't persist until the next day, this might be a sign that you've been visited by a great horned owl. What might have happened is that the bird discovered a skunk, which protected itself in the usual manner. But unfortunately for the skunk, the owl has a poor sense of smell and so carried it off anyway to eat it elsewhere. (34)

If you hear a series of loud, short barks, it means that a squirrel thinks you're intruding on his territory. (125)

There are a number of gadgets on the market that will also help you to find and identify wildlife, everything from night-vision binoculars to motion-sensitive cameras to hearing aids disguised in earmuffs that a catalog promises will have you "hear[ing] a deer walking and grunting and turkey gobbling from half a mile away."

WATCHING FOR DEER

Of all of the mammals that might visit your property, white-tailed deer are probably the ones you're most interested in surveying, either because you like watching or hunting them, or because you want to take precautions if they're coming to your back yard. One way of finding out whether these animals are around is to look for deer trails, which

The shrub's nibbled ends are evidence of deer. (Julie Margerum-Leys)

A whitetail buck is caught at night by a motion-sensitive camera. (Buckshot 35)

will show up as worn-down routes to feeding, bedding or watering areas. (4)

If you are taking a springtime census, hoping to see adult deer, but you find a fawn by itself, never assume that it has been abandoned. What is more likely is that its mother has left it temporarily while she feeds, relying on the fawn's instinct to lie absolutely still so that no one will notice it.

In summer, during the heat of the day, you're more likely to come across whitetails in woods or in the shade of thickets, bedding down on high ridges or sometimes on sunny, south-facing slopes that give them clear views of predators. They move into forest openings when it's dark and cooler. You might come across deer beds, which are usually obvious, packed-down, grassy spots, for deer often return to the same spot. According to Richard Nelson, author of *Heart and Blood*, does with fawns seem to turn up more often where the food is more nutritious, while bucks gravitate to the poorer habitats. (79)

In fall, mid to late September in the upper Midwest and New England, you might find bark rubs on small, supple trees or shrubs. These will be vertical scrapes with shredded bark and exposed wood underneath. (60) One theory is that bucks use these trees to rub the bloody velvet off their antlers; another is that bucks are jousting with a vegetable enemy to practice for the real thing. Watch for evidence of deer around oak and apple trees, in corn and soybean fields. Also watch for white patches where bucks have rubbed the bark of trees with their foreheads to mark the spot with deposits of pheromones. You might see deer scrapes, muddy pawed-up circles about 2 to 3 feet in diameter, where bucks urinate to advertise their availability. In urban areas with lots of deer, it's not unknown for a scrape to occur in the middle of a vegetable garden. Autumn is when you are most likely to spot deer feeding in empty fields at dusk, sometimes in herds of two dozen or more.

After the mating rut, deer will spend more time hiding out in thickets. In winter, before the snow is deep, watch for places where deer have pawed through the snow to plants beneath it. Start looking for evidence of deer yards, where deer are congregating in dense stands of hemlock, spruce, pine, balsam fir or their favorite food tree, arborvitae or northern white cedar. You might see a browse line on trees, an absolutely horizontal line that marks the highest point that deer can reach. Browse lines are not good signs; they mean that deer have become desperate for food. When spring comes in urban areas, you might also see foundation plantings nibbled like unsightly topiary. There will be branches where the snow covered the bushes, and branches where the deer couldn't reach. But in between, the trunks will be bare.

In winter the deer's metabolism slows down and you'll find that they become less easily spooked when you approach them, perhaps because eating less makes their thinking sluggish, perhaps because they sense that bolting often would use up too much energy. (79)

CLUES TO OTHER WILDLIFE

Sometimes you discover animal visitors by examining what they leave behind. You might discover that an owl is making its home on your property if you notice pellets on the ground. These will always be compact packets of the hair, feathers and bones the bird couldn't digest and regurgitated. A barn owl will leave shiny black pellets 1 to 3 inches in length, often beneath the overhead hay track on a hay mow floor. Other owls' pellets are gray. (50)

In winter, deer scat will become shrunken and much drier as the deer are forced to make use of woody browse instead of leafy green plants. (79)

If you find claw marks on beech trees, you'll have found not only evidence of trees that produce heavy crops of nuts, but also evidence of bears. Bears will also leave windrow-like piles of leaves as they forage for acorns. You might also find crushed plants and broken branches, because bears are sloppy eaters. (88)

If someone is nibbling on your valuable plants or trees, examine the leavings before you automatically blame deer. Deer have no upper incisors, so they leave behind torn leaves or stalks with ragged ends. If, on the other hand, you find cleanly cut plant stalks, suspect woodchucks, rabbits or other small mammals. (60)

If you raise poultry and you're losing eggs and birds from areas near the ground, suspect skunks. If chickens are killed and their nests above ground level are destroyed, suspect rats, raccoons, weasels or minks.

If your lawn, your meadow, or your favorite golf course suddenly develops 3- or 4-inch long funnel-like patches of dug-up earth about 3 to 4 inches deep, skunks have probably been busy at night. But if the intruder has produced ridges of dirt with no visible burrow openings, a mole has been at work. (18)

FINDING OTHER MAMMALS

If your land is in an area where beavers might be, look for a telltale dam backing up a stream until the pond behind it is deep enough to provide safe travel to and from a lodge.

Beavers also gnaw down trees to get at the leaves in the canopy. Folk wisdom says that beavers can direct the descent of a tree so that it falls onto the right spot on the dam. According to Benyus, the truth is that the leaves of trees growing in a streamside thicket will develop best where they are open to sunlight, on the side over the water. So when the gnawed tree begins to tilt, it naturally leans toward the heaviest part of its canopy, and falls across the water where the beaver needs it to be.

Beaver dams can cause trout streams to warm. On the other hand, shoreline trees felled by beavers make room for understory shrubs and saplings such as aspen, birch and willow, which many animals snack on. (4)

The dams can serve as bridges for many other animals, and give you a good place to examine their tracks. If there are

Winter is a good time to look for tracks. These are among the most common you might find. (Patricia Gilbert)

no beavers in your neighborhood, other good areas to look for tracks are the edges along rivers, streams or ponds, places where trails intersect, raised knolls and natural springs or seeps. (4) Sometimes a river will provide the only safe travel corridor through farmland or suburbs. Where a river meets any other kind of land, it's an edge, and as with any edge environment, you can expect to see more species of wildlife than you otherwise would. (4)

If you're looking for squirrels, realize that different types have different habits. A fox squirrel will run along the ground to avoid predators, while a gray squirrel will almost always go up a tree first, then it will run or flatten itself onto a branch; nevertheless, its flipping tail will give it away. (17) Gray squirrels prefer dens in trees that they line with shredded bark and plant fibers, but they also build round leaf nests at least 25 feet above the ground in the crotch or branches of trees. Fox squirrels' nests will be at least 30 feet above the ground; they prefer these leaf nests, even where they could easily find tree cavities. (125)

If you live in a northern area of young evergreen forests with dense understories, or young aspen stands, watch for patches worn in vegetation where snowshoe hares have traveled to eat. You might also find a dust bath where small groups of hares have gathered to groom themselves. (128)

As for other mammals, watch for strange caches of dead mice or small birds in a woodpile or hollow tree. You might have found the pantry of a weasel or skunk. These animals sometimes kill more than they can eat all at once, and sometimes store the leftovers. (34)

BENEFITS OF LONG-TERM OBSERVATIONS

Your diary and map should include as much information as you need to give you a base for getting from where you are to where you want to be. If you're up to it, keep the diary not only this first year, but in other years to come. A continuing record will tell you, for instance, just how bad the winter weather has to be before a magnificent but shy male cardinal finally breaks down and pays a visit to your exposed bird feeder. You'll also be more likely to be able to predict the date when the goldfinches will cease to look as dreary as the March weather and start living up to their names.

If you walk the same snow-covered route each winter when the weather is similar, you'll be able to keep track of tracks — how many of each type of animal have come that way. If you do this year after year, you'll begin to see and make use of patterns. For instance, if you see that the raccoon population is picking up in your neighborhood, you might want to plan tighter garbage can and sweet corn security for the next summer.

Write down both rare and common occurrences, because what is ordinary now could turn into a rarity in five years. Your notes might help you discover the change in habitat that caused the disappearance of a formerly common bird or animal, and perhaps even help you to do something about it. (43)

Chapter 4

WILDLIFE IS WHERE YOU FIND IT

So, OK, these theorists talk a lot about observing wildlife in a person's own back yard. But is that really practical? After all, how much wildlife can there be in the back yard, with all the comings and goings of people, with city traffic going by, with lights blazing into the night.

For an answer, let's start with Jesse Ventura, who followed a career in phony wrestling with a few rounds in the political ring. Somebody asked Big Jesse whether he would be interested in running for president when he tired of governing Minnesota. No, he replied, being president is like being in jail.

An intelligent answer. Successful election is followed by at least four years of house arrest, with no chance of visiting the outside world unless accompanied by police motorcycles with screaming sirens and by bands endlessly playing "Hail to the Chief."

But the United States once had a president who was also a naturalist, curious and knowledgeable about wildlife almost to the point of being a scientist, while at the same time a guy who was tough almost beyond the imagination of Big Jesse, a guy who was blind in one eye as the result of a real fight in a boxing ring.

His name was Theodore Roosevelt, and he was a prisoner of the White House from 1901 to 1909.

Roosevelt was a busy president, settling a war between Russia and Japan, digging the Panama Canal, busting up business monopolies, determining a final boundary between Alaska and Canada's Yukon Territory, setting aside 125 million acres in national forests, raising a family of rambunctious children, etc.

In between everything else, he never lost his interest in nature. Unable to roam the wilderness of the Dakota Territory as he had done in his rancher and hunter days, Roosevelt roamed the backyard of that big white house at 1600 Pennsylvania Avenue.

Armed with pencil, notebook, binoculars and that ever-curious mind, Roosevelt recorded every species of mammal and bird, made observations of their behavior, noted the arrivals and departures of migratory songbirds.

Theodore Roosevelt was undoubtedly the last president of the United States who could tell you when the first robin arrived in his back yard in spring and when the last robin left in fall.

Your faithful correspondent makes no claim to be in the same league as Roosevelt, but I have always paid at least some attention to wildlife, especially birds, no matter where I've lived.

Most of those homes have been on small city lots in Sheboygan, Wisconsin, and in Milwaukee County, as well as in cramped student apartments in Madison, Wisconsin, in military barracks in Illinois and Texas, then on somewhat spacious suburban lots.

My dad spent his working life laboring in factories, but that and his lack of much formal education never limited his interest in the wide world. Dinner-table discussions and debates sparked my lifelong interest in politics, history, gardening and wildlife. Dad would remark with equal enthusiasm on what the current president was up to, how that had been shaped by history, and how a killdeer pretending a broken wing had lured him from his tomato patch that day.

My boyhood summer days were spent exploring wooded bluffs along Lake Michigan and watching robins build nests and hunt worms in the small yards of the succession of homes we lived in.

When the U.S. Navy sent me to an air base in Texas, I became fascinated by desert snakes and horny toad lizards and by some of the birds of the desert. Curious as to how fast a jack rabbit could really run, I finally timed one at 35 miles an hour alongside a Navy pickup truck on a back road of the base.

A few years later, a student apartment with a back porch led to observations of little brown bats, indoors and out, along with gray squirrels and nighthawks and other critters of the crowded city.

A big back yard outside of Rockford, Illinois, had shortly before been an alfalfa field, and meadowlarks still sang in their old haunts.

The real estate of a dark and white kingbird overlapped with mine, and the kingbird was right there when I fired up the riding lawn mower. It had quickly learned that the mower kicked up insects as it went. The disoriented insects were easy for a kingbird to catch. So around and around the yard we went, me mowing the grass and the kingbird mowing down the grasshoppers.

Suburban yards in Milwaukee County were near the Lake Michigan shoreline which attracted migrating warblers, Canada geese and hosts of other birds. Birdwatching opportunities were endless.

I first made the acquaintance of a wood thrush, not in the endless forests where I sometimes roamed, but along a rail fence behind a Milwaukee County home. It was under the apple tree in that backyard that I spent some time watching and pondering on cedar waxwings, which came in twittering flocks to feed on the apple blossoms. And it was in that same yard that I learned to tell a ruby-crowned from a yellow-crowned kinglet.

Is it realistic to expect to see and learn anything about wildlife in a back yard?

Of course. Remember that a man named Robert Stroud made an important study of sparrows from the barred window of his prison cell. He became known as The Birdman of Alcatraz, and they made a movie about him.

If he and Theodore Roosevelt could do it, so can you.

— Ron Leys

THINKING SMALL: PLANS AND PROJECTS

What is on the land you own — and what you add to it — will go a long way toward determining the wildlife that visit you. Now that you've explored your own property and the area around it, you should have a good idea of what is already available to attract wildlife, and which kinds of animals, birds and insects you can reasonably expect to see — or see more of — once you've filled in any of the essentials that are in short supply.

In this chapter, we're going to look at things that you can do on a small piece of land with the wildlife that's most likely to live there — birds, butterflies and small mammals. But the same basics for attracting these creatures will apply whether you own half an acre or less, five acres, or an even bigger piece of land.

DIVERSITY, EDGE AND SUCCESSION

"Edge" refers to the area where two different types of habitat meet. It's the place where you're likely to see the most diverse assortment of species. A smaller property is probably located in an area that is already fragmented, broken into lots with houses on them instead of one continuous woodland. That means that you can use the concept of edge to your advantage.

A circle has the least amount of edge. So if you are laying out a plan for your land, a circular lawn or prairie-like open area will offer the fewest opportunities for an edge environment. In a rectangle where the long edge is much longer than the short edge, the edge to area ratio is larger. However, a border that curves in and out not only provides the largest amount of edge, it also looks more natural. (89)

A gradual transition zone between two habitats offers more opportunities for diversity than a sharp break. Even on a small property, you can widen this zone by planting small shrubs in front of tall shrubs, then small trees behind the tall shrubs in order to buffer the abrupt edge. If your neighbor has one or more mature trees just across the property line from your lawn, you might add a transition zone that takes advantage of the tallest layer already in place. (24)

Where lawn meets flower garden, you could plant flowers of increasing heights: low plants like sedum or dwarf marigolds; medium plants like columbine and liatris; tall plants like phlox and yarrow. (24)

One drawback to increasing edge on a small property is that many edge species are adapted to eat a variety of foods — insects, small animals, berries, seeds. Because they are so well-adapted, some edge species become pests when vegetable gardens and garbage cans become fair game. (62)

CHARTING THE FOUR NECESSITIES

It's time to study the map of your property and make some plans. The four things you're trying to locate are space, food, water and cover.

SPACE

On a property as small as a city lot or as big as 3 to 5 acres, some kinds of space may be the most difficult of the four essentials to provide. For instance, in spring, most birds that nest in grasslands need more room than the typical back yard provides. This doesn't mean that you should give up entirely on seeing these birds, for when summer arrives, your property might be visited by young grassland birds that hatched nearby if you provide food or water in a form they like. (107)

PLANNING FOR FOOD

You can plant food for birds, or you can buy it in the store. Seed- and berry-bearing plants, shrubs and trees won't provide food the day you buy them, but once your plantings are established, you won't have to worry about refilling a feeder or keeping it clean so it doesn't spread disease. However, if you insist on instant success, information on feeders follows a little later.

With any luck, there are already things growing on your property that offer food for birds. Use field guides to identify the trees and shrubs that are already growing, then look at Table 4-1 to figure out when each of these makes its food available. Make four lists, label each list with the name of one of the seasons, then match what is already growing to the proper season.

By the way, a shrub or tree labeled "summer" or "fall" does not bear that label because its fruit ripens then. It is a "summer" plant if that is when birds eat the fruit.

If a tree or shrub produces food in fall that birds eat right away, it allows the migrants to build up fat reserves before leaving for the south. Birds that stick around also need to build up fat reserves in order to make it through the winter. American mountain ash, varieties of dogwood and crabapple, Virginia creeper, and winterberry will provide good fall food for birds. (23) Flowering dogwood yields late summer and fall berries that will draw 90 different species of birds including catbirds, mockingbirds, robins, thrushes, woodpeckers, bluebirds and cardinals. (37)

Flowering dogwood provides berries for the birds plus a place to hang bird feeders.

A "winter" plant or tree, on the other hand, often bears fruit that is not immediately attractive, either because it needs one or more frosts to fully ripen (23), or because birds won't eat it until they have absolutely nothing else to eat. A good "winter" plant may also be one whose stems don't break off easily, so its berries remain upright and available above the snow.

Smooth sumac, staghorn sumac, hackberry, hawthorn, eastern red cedar and nannyberry all bear abundant berries that persist into winter. (107) Christmas fern, grapefern and some wood ferns stay green after frost and are, therefore, worthwhile for fall and winter browse. (72)

Don't forget to include weeds on your lists of available foods, if you're willing to leave a place for them on your property. Sometimes weeds will produce more seeds than plants with beautiful large flowers. Some pigweeds have been found to have as many as 100,000 seeds per plant. Ragweed, crabgrass, bristlegrass, goosefoot, doveweed, filaree, smartweed, knotweed, redmaid, tarweed, dock, deervetch — Alexander Martin, Herbert Zim and Arnold Nelson's list in *American Wildlife & Plants* reads like a rogues' gallery. But the seeds are as useful to wildlife as wheat, corn, barley and oats. Moreover, weeds attract insects, which every bird eats at some phase of its life, and some weeds produce usable nectar. And some, like poison ivy, will provide berries for birds. Be aware, though, that some of these plants supply so many seeds that birds will eat only a small fraction and in doing so, may be thinning and reducing competition, which could lead to the enhanced growth of the remaining seeds and the spreading of the seeds of undesirable plants to the yard of a neighbor who may not value birds as much as you do. (72)

Don't get rid of flowering annuals and perennials that have gone to seed, a process known as deadheading, because the seeds that some of those flowers make when left to their own devices are of no less value as bird food than the seeds that you buy in a sack at the grocery store. Cosmos, phlox, liatris and aster and others will provide fall food. (23)

Nuts are valuable for the fat and protein they provide to birds. Acorns are best, with pecans, beechnuts and cultivated walnuts also valuable. (72) However, trees with small, light seeds, like cherry, birch and maple, provide more acceptable food for songbirds than trees like hickory and oak, whose seeds are heavier. (43)

By the time you're done listing your existing perennials, shrubs and trees according to the seasons when they provide food for wildlife, you'll know when your outdoor larder will be well stocked and when it will bare. (37) Obviously, bare is where you should begin to add, especially if the winter food plantings are non-existent. In fact, if you are short of space or money, concentrate on adding winter plants, because winter is the time when the birds that remain in your yard are most in need of food.

PLANNING AND PLANTING

If you decide to sketch your wildlife garden before you plant additions, be sure to draw in trees and shrubs at the size they will reach when they mature. If you do this, you'll probably find that you don't have room for all the things you want, but you will save money because you won't be buying unnecessary plants. (53)

If you were keeping track while the seasons passed, you might have discovered that some of your trees and bushes had lovely flowers, but never developed fruit. Kwanzan cherries, for instance, bear beautiful but sterile flowers. There are also cultivars of virburnums with showy flowers, but no fruit. And hollies and some other shrubs come in male and female versions, but as you might expect, the males don't bear fruit. (37) Such trees and shrubs are great for someone who abhors a messy yard. But such trees won't feed a single bird. So don't add these to your garden, and if they are already growing in a place where room is limited, think about removing them.

If your space is limited, you should also beware of adding plants like the wild plum and chokecherry that spread by suckering. (107) Beware, too, of planting trees where the branches will interfere with power lines or the roots will damage underground drainage tiles. (53)

PLANTING FOR HUMMINGBIRDS

Planting for hummingbirds is a different story because hummingbirds are different from other species. They eat differently, and they act differently. If you think that human beings are mean and vicious, add a hummingbird feeder to the flowers in your yard, fill the feeder, hang it near a convenient window, and watch. For their size, hummingbirds may be the most pugnacious creatures on earth.

During breeding season, male hummingbirds will guard feeding territories and use them to attract females, while the females establish nesting territories and drive other birds away. In her book, *Attracting Birds & Butterflies*, Barbara Ellis suggests that the alternative to all-out war in your yard is to scatter flowers throughout your property or to plant two gardens with flowers that are attractive to hummingbirds, one out of sight of the other. (37)

A hummingbird will patrol a regular route, called a trapline. When it finds something new, it will dive down to investigate, then add it to its trapline in the future. (22) Hummingbirds find red blossoms with tubular or trumpet-shaped flowers, such as those on trumpet vines, especially attractive, but they also drink nectar from flowers of other colors. (23) They'll feed at white and lilac hosta flowers, zinnias, begonias, and sweet williams. (37) These birds seem to prefer plants that are at least two feet tall, and they've been known to visit hanging or potted plants. (107)

A Midwestern landscaper reports that hummingbirds kept getting trapped inside the greenhouse where she was raising petunias for sale. The cure? A pot of petunias at the doorway pointed

A hanging plant can become a hummingbird feeder, especially if the flowers in it are red. (Sandra Stark)

the way from the indoor petunias toward the outdoors where the birds belonged.

Hummingbirds are also meat eaters, deriving their protein from weevils, gnats, aphids, small beetles, mosquitoes, flying ants, leafhoppers, flies and daddy longlegs. Ruby throats have been seen picking insects out of spider webs. (37) It goes without saying, then, that insecticide is not going to lure any hummingbirds to your garden.

PROVIDING SNAGS

To some birds, "food" means small creatures like mice, rabbits, or even other birds, while other birds define "food" as insects. Snags are gifts to predators like these. The insects that thrive in snags or dying trees will provide insect-eaters with protein. They will also provide perches from which predators can scout for their prey.

If there is a snag already on your property, keep it. If not, you might want to provide one by killing a tree six inches in diameter or a little larger. All it takes is an ax — use it to cut away a continuous band through the bark and into the cambium layer beneath it near the base of the tree. (28) If you or a neighbor feels that the snag is unsightly, you can disguise it with a trumpet vine and attract hummingbirds at the same time you provide for woodpeckers. (37)

Red-tailed hawks, kestrels (sparrow hawks) and other raptors that forage or nest in open country appreciate tall perches. Eastern phoebes, eastern kingbirds, northern mockingbirds, song sparrows, eastern meadowlarks and other singing birds and birds that prey on flying insects make use of perches like fence posts that are less than 10 feet tall.

In one densely populated neighborhood of Philadelphia rowhouses, northern mockingbirds like to perch on rooftop television antennas. They'll produce pretty good imitations of a robin's evening song and not-very-credible cardinal whistles, even though they are only half a block from a very busy commercial avenue.

UNDERSTANDING BUTTERFLIES

Butterflies can make use of an even smaller space than birds if the right kind of food supply is there. Even one butterfly bush, or a patch of red valerian, or a small patch of lavender will be enough to attract some butterflies, if your winters allow you to grow one of these plants.

A butterfly bush is the magnet its name implies.

To understand why a butterfly garden needs certain patterns and plantings and auxiliary equipment, you need to understand a few things about a butterfly's life cycle, eyesight and habits.

When a butterfly lays her eggs, she usually glues one egg to each site. She chooses these sites with much more care than she uses when searching out nectar plants for herself: Table 4-2 lists many more nectar plants than larval food plants for each butterfly on the table.

A black swallowtail will land on a carrot leaf, and after identifying it with the sensing devices in her feet, will deposit an egg. If she finds herself on a tansy or yarrow plant, which has leaves that look something like the carrot's, her feet will tell her that she's in the wrong place, and she'll fly on. All this care is necessary because her eggs are abandoned as soon as they are laid; therefore, she has to think ahead, considering what her offspring will be willing to eat as they fend for themselves.

Eventually, a tiny caterpillar will eat its way out of each egg. At this point in its life cycle, "a larva is an eating machine," writes Dave Winter. It will devour leaves of the plant on which it finds itself, and stop eating only at those times when it is shedding its old, too-small skin. (121)

For most caterpillars, there will be five instars, or growth periods, separated by shedding the outgrown outer layer. This phase will last three to six weeks if conditions are right.

During its last larval stage, the caterpillar will look for a place where it has a good chance of surviving after its skin hardens into a pupa. In this sack — called a chrysalis when it belongs to a butterfly — the caterpillar will transform itself into its final life stage. It might remain on the same plant, moving just a few inches to a new leaf or stem, or it might crawl hundreds of feet, seeking a patch of soft soil where it can bury itself. (121) Temperature and the time of year control the length of this stage; some butterflies even winter over as pupae.

When the pupa bursts open, the adult climbs out and seeks a sheltered spot where it can let its soft, floppy wings hang down. As it pumps fluid into the ribs of those wings, they become longer, spread out, and form the framework for the membrane that spans the ribs. Butterflies of smaller species need about 10 or 20 minutes for their wings to harden, while large moths need several hours. The colorful patterns that help us and predators tell one butterfly from another are actually tiny scales laid on top of the membrane like shingles. (121)

Even a small urban herb garden will attract butterflies if it has the right stuff. This one includes lavender and hyssop. (Sandra Stark)

The right trees can also be magnets for butterflies and birds. (Sandra Stark)

When they are fully prepared, all butterflies will go off looking for nectar and the males will start looking for females. Nectar provides butterflies with the carbohydrates that give them energy for flight. Different flowers contain one or more of the three kinds of carbohydrates, sucrose (as found in cane sugar), fructose (fruit sugar, the sweetest of the three), and glucose (as in corn or grape sugar). Butterflies can sense what is available through chemical receptors on their tongues, antennae and feet. (101)

Butterflies can see more than we can at the ultraviolet end of the spectrum, but less at the red end. In fact, some butterfly species don't see red at all, leading to a theory that the new shoots and leaves of many plants are colored red so that a butterfly cruising around, looking for a place to lay eggs, misses these plants, saving them from being devoured by caterpillars. (100)

There's evidence that butterflies do not have good distance vision. One study found that some flowers that seem to have solid-colored petals in normal daylight show guidelines when viewed under ultraviolet light. These guidelines point the way to the source of the flower's nectar as surely as runway lights point the way to an airport. Flowers that attract butterflies have these guidelines, hidden to the human eye but readily apparent to butterflies. However, the types of flowers that are pollinated by hummingbirds, whose eyesight is known to be keen, do not have guidelines. (100)

On a dry day, creating mud will attract butterflies that puddle.

As the puddle becomes mud, a regatta of sulphurs forms.

Additional evidence about butterflies' poor eyesight comes from the fact that when a male butterfly is trying to mate, he patrols or flies over an area where females might be feeding or laying eggs. However, when he thinks he's found Ms. Right, it's necessary for him to swoop down and examine her more closely before beginning his courting ritual. (64) A male cloudless sulphur butterfly on patrol will inevitably investigate a yellow paper attached to a stick. A male fritillary will fly toward an orange tossed into the air. (121)

Butterflies are cold-blooded, which means that their body temperatures will be very close to the temperature of the environment they find themselves in. However, for a butterfly to fly efficiently, its wing muscles must have an internal temperature of 75 to 110 degrees. So what is a butterfly to do? For most species, the answer is: bask. Basking for many species means sprawling with wings open flat on a warm, flat surface for a minute or two as the muscles absorb heat both from the sun and from the sun-warmed soil. After the body temperature reaches flight level, the sun's heat plus the heat produced by flying will be enough to keep the butterfly warm. But if a cloud appears, then it will be forced to settle to earth again. (121) Butterflies that bask using this method will often have black bodies and dark areas on their wings to make the best use of the sun's warmth.

Sulphurs and satyrs use a method called lateral basking. That is, they fold their wings together and turn to face the sun, catching perpendicular rays. (64) An alternative to basking is shivering, contracting the flight muscles rapidly to vibrate the wings without actually flapping them. (121)

One more butterfly behavior that you will want to take into account is puddling, often done in groups. A group of puddling suphur butterflies looks like a miniature sailboat regatta. Usually more males than females will appear at a patch of mud in order to extract nutrients. This seems to help the males produce pheromones to attract females. (64) Butterflies are often seen using the damp soil or sand at the edges of ponds or streams. These patches contain salts and minerals that have leached from dead animals or animal manure. (11)

RESEARCHING YOUR NEIGHBORHOOD BUTTERFLIES

Knowing something about butterflies' habits will help you as you do some research before planting your butterfly garden. For instance, knowing about butterflies' need for warmth tells you that it's best to take a butterfly census on a sunny day. As you walk around your neighborhood, notice which butterflies are flying in the woods, in vacant open areas, in railroad right-of-ways. (10)

Butterflies are most easily watched and identified while they are drinking nectar from flowers. Sometimes they will get so involved with nectaring that they don't even realize you're close. The butterflies basking in a sunny spot will also make identification easier, giving you a window of a minute or two while they lie as flat and still as if they were pinned to a board.

Because of its poor eyesight, a butterfly might mistake you for a potential mate, a potential enemy or a flower if you're wearing bright clothes or a bright bandanna. Your clothing might make the butterfly come closer. If it's a hot day and your clothes are sweaty, you might be especially attractive. (96)

At least as important as identifying the butterfly is identifying where it is nectaring and laying its eggs, because this will give you a good idea of what to plant in your own garden. (10)

After studying your neighborhood, study your own yard. How much sun reaches the area that you want to use? Are some of the plants or bushes providing so much shade that they'll have to be moved? Butterflies need sunshine for warmth, but they also feel more comfortable when they don't have competition from a stiff breeze. So think about whether you need to plant or build windbreaks.

Something to think about when you're deciding on the size of your garden is the amount of maintenance you're willing to put into it. Remember that chemical sprays do not help to attract butterflies. (10)

Monarda (bee balm) (Sandra Stark)

When choosing the array of plants for your garden, consider that butterflies are much more discriminating about where they lay their eggs — larval host plants — than they are about where they get their nectar. This means that to be successful, your garden should provide food for both butterflies and the caterpillars that precede them.

In their article, "Butterfly Garden Design," Mary Booth and Melody Mackey Allen advise that you start by choosing a small bed protected from the wind where you plant three to four species of perennial nectar plants with a range of blooming times. Bee balm, butterfly bush and summer phlox might be good for a start. (10) Bee balm is in the mint family and will spread, so you might want to take the bottom out of a plastic bucket, dig a hole big enough to accommodate the bucket, sink it in the ground, then plant the bee balm inside the bucket to retard the spread of its roots. Be aware, though, that after several years, the only growth may be around the edges of the bucket, so you'll have to dig up the bee balm and re-plant it. In areas with very cold winters, butterfly bush can be a little tricky to get going and keep going. It will die back to the ground each winter, and might not come back at all in spring. However, as the name suggests, it can be worth the effort.

After planting a few perennials in the first year, add plants in the second year to extend the blooming season, plus one or two species of annuals to feed the caterpillars. The third year, increase the selection of nectar plants again, then try more unusual native larval and nectar plants from a local native plant nursery. (10)

The purpose of the host plants in your garden is to give caterpillars something to devour, so you shouldn't get upset if these food plants are getting eaten. Of course, some of these plants, including herbs like parsley, fennel, and dill, and flowers like lupine and sweet pea are of use to garden owners as well. If you expect to have some leftovers for yourself, be sure to plant enough for everybody. (23, 107)

Some caterpillars come equipped with forked spines or hairs that may be for show or may sting. If you decide to move this type of caterpillar because it is eating too much of a plant, especially one in clear view, wear gloves to protect yourself as you move the larva to a less conspicuous place. (3)

Single, open-faced flowers give butterflies the best landing pad. (Sandra Stark)

If you have a weedy corner, don't be in a hurry to clean it up. Burdock is a host plant to painted ladies, and plaintains host buckeyes. Variegated fritilleries and Milbert's tortoiseshells, question marks and red admirals will lay their eggs on nettles. In fact, the easiest and cheapest way to establish a nursery for butterflies is to simply not mow an area of your lawn and let the weeds accumulate. (37)

When caterpillars become butterflies, they lose their mouth parts and rely on nectar alone for sustenance. As they gather that nectar, butterflies will do your garden a service: they will also inadvertently gather pollen, pollinate other plants as they seek nectar from those flowers, and secure the future of your garden.

BUTTERFLY FOOD

Butterflies like to visit a variety of species because each has a different percentage of the high energy sugars and lipids that they need. (107) But whatever you plant, you should be grouping flowers of similar colors together in order to give the possibly-nearsighted butterflies as big a target as possible. (64)

What should you plant or encourage? Once again, a weed is at the top of the list. In *Notes from a Butterfly Gardener*, Jo Brewer calls milkweed "one of the most important of all butterfly nectar plants." She has counted 18 species of butterflies nectaring at milkweeds, and she points out that the monarch female uses it as a place to lay her eggs. (11) In fact, the attractive plant known as butterfly weed for obvious reasons is also a variety of milkweed.

Ellis suggests leaving some room in your garden for Queen Anne's lace, black-eyed Susans, wild asters, goldenrod, milkweed and yarrow. Or, if you have a shady garden that's moist or wet, yellow and orange-flowered jewelweed. (37)

Table 4-1 lists trees that are useful to butterflies. Table 4-3 tells you which plants will attract butterflies. Remember, though, that sometimes a plant that a book identifies as supposedly good for butterflies will get no attention in your personal garden. The problem may lie with the preferences of your local butterflies. (64) Sometimes, when you're on vacation in a different part of the country, you'll notice butterflies flocking to a particular flower. At home, you have the same species of butterfly, so when you get home, you plant the same flower, because the plant-hardiness map assures you that you can. Then your butterflies ignore it, because to them it's a foreign species that they simply don't recognize. The moral of the story: buy plants that are native to your area or transplant them from nearby areas that are being bulldozed and converted. (11)

Judging by the flowers they are attracted to, butterflies prefer strongly perfumed plants to lightly scented ones. (101) Plant large, single-blossomed, upright flowers because single blossoms are easier for butterflies to land on and have richer nectar that is easier for them to get to. (23)

Various butterflies will get food from other sources as well. They will feed on the juices of dead rabbits, snakes and frogs, though you'd probably draw the line at providing these. However, some species love overripe fruit like mashed bananas. You can experiment with combining fermented fruit with wine or beer and a little honey or sugar. Pour some on a sandy spot, keep it damp, and see who comes, suggests Brewer. (11)

BEYOND FOOD

Unlike birds, butterflies don't need supplies of water. All they need is a damp spot to draw an occasional sip, or a little mud for puddling. On a hot, sunny day, the swallowtail that seems to be heading for your newly watered petunias may seek out, instead, the damp mud at the flowers' base. If your soil tends to dry out very quickly, you might think about sinking a small bucket into your garden, filling it with sand, then keeping the sand moist.

As for basking, set a flat, light-colored stone where the morning sun will warm it most efficiently. This will give butterflies a place to spread out and warm up their muscles at the time of day when they most need this kind of help (107) as well as a place where you will almost always find them when the weather is right.

You can shelter butterflies from the wind, but for the most part, you can't do much about giving them shelter from predators. Actually, both butterflies and caterpillars have managed this on their own.

Larvae tend to be masters of disguise. The best look exactly like bird droppings. As another form of protection, some caterpillars can isolate and concentrate chemicals on their bodies that, when transferred during pupation, make the resulting butterflies distasteful to birds. That's one function of their bright coloring: warning off potential predators. (64) The monarch butterfly is especially good at using the chemicals it finds on milkweed to protect itself from its bird enemies by making them so sick they puke. The queen and viceroy butterflies have adapted to take advantage of the monarchs' bad taste: their coloring has come to mimic the monarchs', so birds avoid them too, even though they would make a tasty meal. (69)

Some butterflies have developed eye spots on their wings that seem to startle birds, while in some, the bright coloring on the tops of the wings is matched by dull coloring on the bottoms, allowing them to fold their wings together, hang from twigs, and look exactly like leaves. (100)

Despite butterflies' mastery of disguise, collectors frequently notice the v-shaped marks caused by missing scales on their wings or v-shaped tears at the edges of their wing membranes. These indicate that birds have attacked the butterflies unsuccessfully. (100)

It stands to reason, then, that if you are taking the trouble to lure butterflies to your garden by planting nectar-producing flowers and trees and food for the larvae to devour, setting out flat stones for basking and maintaining mud for sipping and puddling, you should not be hanging bird feeders directly above your butterfly garden. (107)

Will all of this work guarantee success?

Some species of butterflies will remain in a bog or some other particularly attractive habitat, especially if other areas like it are far away. Other species with less specific requirements may move a mile or more from day to day, always on the lookout for better nectar. Chances are that the butterfly you see today in your garden is not the same one you saw yesterday, though missing wing parts and scales will help you identify insects if you want to take the trouble. (121)

But chances are also good that the butterflies will visit if you prepare even the smallest table for them.

PLANNING FOR COVER

How many trees are on the map you made of your property? How are they arranged? A single specimen tree may be beautiful to your eye, as a picture on the wall in an art museum is beautiful, but it does not offer much in the way of cover to birds who are trying to shelter in it, especially when a bird of prey also finds that the tree is a handy place to sit.

The truth is that blocks or clumps of trees or shrubs are more useful to wildlife. If you are starting from scratch, plan clumps of species that need similar amounts of water. If you have limited space and/or money, make your first purchases the trees that will take the longest to mature. Or buy trees and shrubs that provide both food and cover for birds. (107) Table 4-1 will help you identify trees and shrubs by the way wildlife uses them.

On a property of five acres or less, concentrate on evergreens to shelter wildlife from winter winds and heavy snowfall. Plant shrubs that provide fruit at different times of year so wildlife has a continuous supply of food. Larger properties will have the space for a shelterbelt, an L-shaped belt several trees wide that protects a house and wildlife from the prevailing winds. See Chapter Five if you have the room to plant one.

If there is not enough room for a shelterbelt on your property, plant clumps of evergreens like eastern red cedar with a ring of fruiting shrubs around the outside edges. (107)

Besides giving wildlife winter cover, a row of evergreens on a small lot will give you some privacy. Hedges planted on your lot line will provide a living fence that is far more useful than anything you can buy in a lumber yard. Choose the right shrubs and plant them thickly and you'll be giving birds food and a place to nest. (53) For instance, shrub roses and hawthorns provide good nest sites because their thorns offer added protection from predators. (37)

If your property allows room for a small grove of trees, diversity is important in order to provide cover for a variety of species. Clumps of evergreens should be mixed in with hardwoods to provide winter cover for birds.

If mourning doves are in your area, they might also be in your yard, since they don't make use of deep woods. They are generalists about nesting cover, as long as you stick to evergreens. Norway spruce, white pine, red cedar, Scotch pine, Virginia pine, Japanese larch and hemlock all give mourning doves a place to nest. Doves seem to prefer trees whose lower branches droop to the ground instead of trees that look neater but provide less cover. (43)

PROVIDING COVER

Brush helps wildlife in a city yard as much as it helps in a young forest. If nothing but grass is flourishing under your trees, create some animal shelter by removing all of the grass, then underplanting the taller trees with shade-tolerant shrubs and short trees. Depending on the part of the country in which you live, such shade-tolerant plantings might include dogwoods, azaleas, rhododendron, shade-tolerant viburnums, hollies, blueberries, Oregon grape holly and spicebush. (37)

At the edge of the trees, you could put in crabapples, or blackberries and raspberries (111). The berries' thick canes will do double duty by providing food as well as shelter. Blackberries, raspberries and others like hawthorn and witch hazel, do best in direct sunlight. (43) In one study, berry bushes in full sun provided 30 times more fruit than the same plants growing in wooded shade. (4)

If your trees start shading these shrubs, think about trimming back the trees' canopies. Mowing the canes will stimulate sprouting and get rid of competing plants. (43) If you only have a small patch and you expect to eat some berries yourself, you're probably going to have to cover some of the canes with netting. On the other hand, if your space is less limited, you might start a thicket for the birds in an out-of-the-way corner. You'll probably have enough berries for everybody in not very many years, because where these canes grow, they tend to spread very quickly.

To break up heavy soil and improve your plants' chances of thriving, use shredded leaves as mulch or turn the leaves into your soil. Shredding speeds up the process of breaking down the leaves and helps keep them from matting. If you don't own a shredder, run your lawnmower over piles of leaves. (111)

A snag or a living tree with a hole in its side will provide nesting and denning cover. Trees that rot most quickly will produce den sites most quickly. Rot happens faster in hardwood trees than it does in evergreens. On the other hand, a tree whose core rots faster will not remain standing and useful for as long a time. (110)

As you might expect, animals and birds have definite preferences for the trees where they will nest. Red-breasted and white-breasted nuthatches don't create their own holes, but they will use existing cavities in living trees with either central decay or broken tops. A hairy woodpecker drills its own nest in any living or dead tree except a dead

Holes in a dead tree may provide dens for wildlife.

evergreen. A winter wren will use an existing den only in a dead evergreen. Red, gray and flying squirrels, and porcupines will all use existing dens in either hollow trees or living trees with broken tops and limbs. But gray foxes, black bears and raccoons will only den in hollow trees. (110)

NATIVE PLANTS FOR COVER AND FOOD

Tearing out an urban or suburban lawn and replacing bluegrass with something else has become an option for some city dwellers because native plants are often better adapted to withstand local extremes of weather and attacks by local insects, while native grasses take less water and care than lawns. (107) Butterflies are also more likely to recognize plants if they are native to the area where they are seeking nectar or a place to lay their eggs.

Native plants can share a standard garden, or you can create a prairie. The original prairies were large, open expanses that at first glance looked like nothing but tall stands of native grasses. However, closer inspection revealed prairie wildflowers peeking through. Urban prairies tend to emphasize flowers for a variety of reasons; many of these flower-prairies will cover an area as wide as the house's former lawn.

Because prairie plants evolved where frequent fires were a fact of life, more than 60 percent of each of these plants is located underground and is well protected by the soil that covers it. Prairie plants have developed a way of making food that allows them to grow actively when it is extremely hot. Their version of photosynthesis takes much less water. Some prairie plants lose less water because of their small or finely cut leaves or smaller pore openings. Some plants have hairy surfaces to reduce the flow of hot air that causes evaporation; some have leaves close to the ground where the flow of air is reduced. Some have deep taproots or wide-spreading fibrous roots to make maximum use of the water that is available. (2) In fact, once root systems are established, too much water will result in plants that are too tall and spindly. (75)

Once established, prairie grasses work with the flowers to keep weeds

Black-eyed susan

Echinacea (purple coneflower) (Sandra Stark)

away. Because the grasses have relatively shallow roots, they don't compete with the flowers' roots — some as long as 10 feet — but they do cover the bare ground, cutting down on the available space for weeds to grow. However, the owner of a small prairie who wants to emphasize flowers over grass should plant the shorter grasses that grow in clumps, like little bluestem, sideoats grama, or prairie dropseed, in order to leave room for flowers. The tall grasses like big bluestem form dense sod and are more likely to restrict the growth of flowers. (94)

A site as small as an acre or less will provide a sampler of prairie plants, but if you want to create several kinds of prairies and increase wildlife habitat, a larger area is more likely to guarantee success. On the other hand, if you own a large property, but you don't have a lot of time to spend when the prairie is new, your chances of success are better on a smaller plot. (75)

PREPARING YOUR PRAIRIE

Though some of the gardeners you will meet later in this book have had luck merely plunking prairie plants into existing sod, most experts agree that preparing the entire site before you plant is the best idea. One option is to smother the weeds in the plot for a full growing season by covering the ground with black plastic or other impenetrable materials. Another is to use a broad-spectrum, non-persistent herbicide; glyphosate (Roundup, Ranger or Kleenup) is the best choice. Or, for a larger area, cultivate with a sod-cutter, rototiller or tractor-mounted implement to break up the top three inches of sod and soil. (94)

While your soil is preparing itself, it's time to plan what you want to plant, and in what form.

If you're planning a small prairie, like one as small as the corner of a yard, you'll have better results if you install plants instead of seeds because young plants that have gotten a head start are much easier to spot among weed seedlings. Besides, prairie seeds have a habit of growing somewhere besides the area where they've been planted. (75) Another benefit of plants is that, with care, transplants often bloom the first year, while flowers planted from seed usually don't bloom until the third year. However, if you're planning a prairie of more than 1,000 square feet, seeds will be much more economical. (94)

If you choose plants, plan on mulching with three or four inches of clean straw to hold down the weeds. (94) Try to transplant in spring, when rain is the most abundant. Most prairie plants will not survive if they're transplanted in fall because ground heaves in late winter will expose the plants' crowns and the plants will die. (75)

The more habitats you have on your prairie, the more species of wildlife you will attract. According to one supplier of prairie plants, another benefit of variety is that people who install mass plantings of only one or two species of flowers have more weed problems than gardeners who choose variety. Once your prairie is established, you can create diverse habitats by dividing the prairie in half or in thirds, then burning or mowing each section only once every two or three years. This will automatically produce changing patterns within the same planting. In addition, the untouched section will protect overwintering butterfly chrysalises and give birds more nesting cover. (94)

If you want to attract butterflies and other insects, plan to include coneflowers, blazing stars and goldenrods. (75)

"BORROWING" PRAIRIE SEEDS AND PLANTS

Instead of buying seeds or plants, you can collect them from lands where they are already growing, but always ask the landowner's permission before doing this. When you collect, always leave more than you take so that the land can replenish itself rapidly. Collecting seeds rather than taking plants will obviously leave more for the future.

The exception is land that is about to be developed. Real estate brokers' signs or surveying markers are good clues that development is about to scrape away all of the existing greenery. Again, ask the developer's permission before you go to work digging up flowers, other plants like mosses or sedges, or shrubs and small trees.

To help reduce the shock of transplanting, cut back the top third of each plant before you dig it up. (2) If you're moving a small tree or shrub, you can reduce transplant shock by root pruning the year before the move — assuming you have enough time. Before growing season, dig around the plant to the depth of the shovel blade, two or three feet from the trunk or stem. This will cut the shallower roots and force the plant to form a tight ball of roots for easy moving. At the same time, prune about half to two-thirds of the branches so leaf and root surfaces balance. The next year, move the tree or shrub in fall if you live in a milder climate, or in colder climates move it in the early spring, after the ground thaws but before your tree leafs out. Cover a bare-root deciduous tree with a damp burlap sack or sphagnum moss until you re-plant it. Evergreens' roots should be lifted with a ball of dirt surrounding them. When you put the shrub or tree back in the ground, be sure it is planted to the same depth and no air pockets surround it. (42)

When you're done, be sure to thank the developer — you might want to do some rescue work on some future project of his. (2)

If seeds are your choice, Barbara Ellis advises that you avoid buying cheaper prairie seed mixes or mixes that promise you instant success because they will contain mostly annuals. Remember that what you are trying for is a permanent installation that within two or three years will save you a lot of time and water. (37)

Try to buy seeds that are grown in your U.S. Department of Agriculture (USDA) hardiness zone, preferably within 50 miles of your home. (107) This is important even if re-creating a purely native prairie is not one of your goals, not so much to be sure that your plants will survive, as to guarantee that they don't survive too well. An exotic plant lacking in natural enemies might take off in more directions than you had planned. If a plant, shrub or tree is native to your area, it probably has enough natural enemies to keep it in check. If it's not native, contact someone in your state's university extension service for a list of ecologically invasive species, but be aware that the department may only be concentrating on species that are causing problems to farmers. Your state's natural resources agency may also have an expert in the species that are invasive in your area.

PLANTING YOUR PRAIRIE

It's best to plant your prairie in spring or early summer, when you can control the weeds better before you plant. If you have dry soil, you'll be taking advantage of the time of year when rain is usually the most abundant. If you have clay soil, you'll be giving the seedlings a chance to put down roots before the clay dries so hard that root growth will be restricted. (94)

Just before you are ready to plant, lightly cultivate only the upper 6 to 8 inches of soil. Thorough cultivation will create a bed of loose soil and prairie plants need the soil packed tightly around their roots. However, tight clay soil may need sand and peat tilled in and thoroughly incorporated before you add the prairie plants. (75)

Hand seeding will work on a small prairie. Mix the seeds with a lot of sawdust, peat moss, vermiculite or other inert, lightweight material. For a planting that's 1,000 square feet in size, a bushel of this material will be enough. Walking in one direction — north to south, for example — broadcast half of the material. Then turn and walk in the direction that's perpendicular to your first seeding — east to west, in this case.

Rake the soil so that the seeds are covered by 1/8 to 1/4 of an inch of dirt, then run truck or tractor wheels over the soil to compact it. If the area is small enough, you can even stomp the ground with your feet. The object is to remove air pockets around the seeds to cut down on water evaporation. A larger area might require that you rent a seeder that can handle the lightweight prairie seed, then roll the dirt to firm it. (94)

The news on weeding during the first year is that you should do as little as possible. Prairie plants will be spending most of their energy growing deep roots, so what you see above the ground is very small and hard to identify. Not only might you miss the seedlings you want to keep, you might also expose additional weed seeds by shifting dirt. Your best bet is to pull weeds only while they're young and small and use a pruning shears on taller weeds, then remove seed-bearing stems from the area right after they're cut. (94)

To suppress weeds in the second year, mow everything to the ground, then rake off all of the cuttings, if your area is small enough. According to Prairie Nursery's directions, burning this early in your prairie's life does encourage dormant prairie seed to grow, but by exposing the soil it can also lead to more weed growth. If weeds are still a problem, mow a second time in late spring or early summer to about one foot, while the weeds are in full bloom. This should kill, or at least discourage, most of the weeds.

The downside of establishing a prairie for the owner of a small property is that a true prairie needs to be burned periodically to maintain the vigor of the plants. (2) If periodic burning is not an option where you live, but you like the idea of replacing your lawn with native plants, you could plant native wildflowers in a much thicker concentration than would appear on a prairie and mow periodically in mid-spring, removing the clippings to expose the soil and encourage the growth of heat-loving prairie plants. (94)

MAKING YOUR NEIGHBORS HAPPY

A stretch of grasses and wildflowers, especially if some of them are 6 feet tall, is a change from a mowed lawn that is shocking to some. If you live in an urban or suburban area where mowed lawns are the norm, and you're thinking about substituting a prairie or wildflower garden, there are a number of places to go for advice, including Wild Ones, an organization that encourages the substitution of native wildflowers for grass lawns. Bret Rappaport of Wild Ones has some suggestions that he condenses into the acronym BRASH — Border, Recognize, Advertise, Start small, Humanize.

People will accept something if they sense it has a purpose, but they will reject exactly the same thing if they feel that there is no plan behind it, Rappaport says. So put a border around your garden. It could be a strip of lawn, a hedge or a

Tall prairie flowers may seem like weeds to an untrained eye. (Sandra Stark)

Manmade touches can be reassuring to someone unfamiliar with a prairie. (Sandra Stark)

fence, some low native plants or a path. A border not only shows that you're planning, but it will keep tall native plants from falling over into your neighbor's yard, and will give oncoming automobile and pedestrian traffic a clear view.

No matter how convinced you are that native plants are the way to go, recognize that your neighbor also has the right to do whatever he wants in his yard. Don't preach or disparage his landscaping efforts.

Advertise. Let your neighbors know your reasons for tearing up your lawn. Before you begin, tell your plans to your neighbors and your local officials. After you've started your garden, put up a sign. Again, this gives people the idea that you have a plan, and that your yard should be admired, not condemned.

Rappaport also advises would-be native gardeners to start small. Changing from grass to native plants a small patch at a time will spread out the expense and work, allow you to learn from small mistakes rather than from a yard full of them, and allow your neighbors to get used to the change more gradually.

Lastly, he urges you to humanize your natural plot by adding touches that have obviously been created by human beings. A path running across the garden, a bench where people can sit to admire what you've done, bird feeders or birdbaths, even a sundial or a small, picturesque relic of a bygone time, like a wagon wheel. Again, such additions show your neighbors that there is a plan behind this yard. They also make your natural landscape less threatening to people who are not used to uncut grasses sprinkled with flowers. (2)

William McClain adds that when plants are too close together, they may be perceived as weeds. He advises leaving enough space between plants so that most sides of the plant and the area surrounding it can be seen. Like Rappaport, he also advises gradually increasing the height of the plants in your small prairie, though for a different reason: plants of mixed heights say "weeds" to the critical.(75)

Bird Feeders and Other Shortcuts for Attracting Birds

Planting to provide food or cover is an ideal way to improve your property for the birds you might bring to it. But such projects take time — sometimes several years — before you will see results. They also take more space than you might have.

If this is the case on your property, there are other options, like feeders, nest boxes and brush piles.

Bird feeders have advantages and disadvantages. They can supplement birds' diets during a lean time of year and bring birds close to you for ease and predictability of viewing. On the other hand, they can encourage sick birds to mingle with healthy ones, and they concentrate potential prey in a place where predators know they can get an easy meal. (107)

Some feel that feeders make birds dependent on a food supply that will be uncertain if the caretaker quits; others point out that the most successful birds in the wild are the ones that can find another source of food quickly if the first source dries up. What is indisputable, though, is that in cold weather, birds that aren't able to become torpid (allow their body temperature to drop to conserve energy) have to eat continuously through the short winter days so that they have reserve energy to keep themselves warm through the long winter nights. For instance, a chickadee must spend about 20 times more time feeding when winter temperatures are low as it does in spring when things are warming up. (35)

You can feed all year long if you want a constant stream of birds at your window, but they really only need this artificial source of food in late fall, winter and early spring, when natural sources of fruit, seeds and insects are scarce to nonexistent, times when it's beneficial to expend fewer precious calories getting to a feeder than wandering to search for leaner pickings.

This doesn't mean that you have to provide for every bird that shows up. Instead, set a bird food budget and parcel out your food accordingly. This is better than starting to feed in December, trying to keep up with the feeding of a growing number of birds, then running out of money before spring. (34)

DANGERS OF BIRD FEEDERS

If possible, your feeding station should be in a warmer location in winter and a more shaded area in summer. It should also be in an open space at least 10 or 15 feet from places where predators can hide so that songbirds have a clear view of the territory and the dangers that might lurk there. (107)

If your lot is very small and your feeder has to be very close to trees or bushes, it might well attract predators. If this is happening, surround the feeder with two-inch

Bird feeders come in may shapes and sizes. Here are a few: (1) tube feeder with perches below the portals (2) suet feeder (3) tray feeder for ground-feeding birds (4) tube feeder with perches above the portals to encourage agile goldfinches (5) fly-through feeder (6) fully enclosed self-feeder. (Patricia Gilbert)

by four-inch wire mesh three feet high to keep raptors and cats from taking advantage of the situation. (45)

No matter how far your feeder is from cover, you're probably going to want it where you can watch it from the comfort of your house. But this presents its own possibility for danger: that window you're looking through is invisible to the birds, which might crash against it and die before they know it's there. Sometimes the danger will come from a raptor or crow that has learned that if it flies toward a group of birds at a feeder near a picture window, some of those birds will knock against the window and the predator will seize the opportunity, plus the stunned bird. (45)

One solution is to do something to make the window more visible. You could install plastic garden netting on a frame and put it over the window, or for something really effective but ugly, paste on strips of colored tape. (107) Several writers report that falcon silhouettes do not seem to reduce bird strikes even though they make the windows more visible.

The easiest solution of all, perhaps, is to move the feeder so it's only a foot or two from the window. Then, if the birds are flushed by predators, they won't develop enough momentum to crash hard against the glass. (45)

TYPES OF BIRD FEEDERS

The type of feeder you use depends on what you mean to feed from it, who you mean to feed, and how often you want to re-fill.

The simplest type of feeder is an open tray that can be screwed onto a log, post or railing, hung from a tree branch, or built with feet to stand on the ground. This can be used to feed pieces of fruit or bread as well as various types of seeds. An open feeder can also be used to offer mealworms in a smaller, shallow container. Open feeders should not be filled with any more food than the birds will consume in a day or two. Ideally, such a feeder should have a few drain holes in the bottom, and a gap in one side where moldy seeds can be scraped out. Waste seed that has fallen and remained uneaten should be removed every once in a while so that mice, rats and voles are not attracted. In addition, sunflower seed hulls that drop should be cleaned up, but don't throw them where you expect other seeds to grow or on compost piles, because the hulls contain a chemical that inhibits plant growth.

The benefit to this type of feeder is that birds have a clear view of predators because there are no sides to block their view. The drawback is that it is completely open to the weather, so rain wets the seeds, which may develop harmful mold. (45)

A slightly better option is a fly-through feeder, a tray feeder with a gable roof attached to corner posts about 10 inches high. Like a tray feeder, this offers birds a clear view of predators, but the roof keeps the seeds drier longer.

Many people choose a self-feeder, a tray feeder with a roof and plastic or glass sides that attracts all sorts of songbirds. It needs to be refilled less often than a feeder with no sides, but it can attract squirrels and undesirable birds. (45) In fact, as the seed level goes down inside the walls, house sparrows can get inside the glass to gorge themselves on the seed that they're sitting on without fear of competition.

Glass or plastic tubes with openings of various sizes are also popular feeders.

Feeders with large ports accommodate sunflower seeds and mixes that contain them. The best of the cylinder or tube feeders have metal grommets around the openings to cut down on squirrel damage. Small birds, including purple finches, goldfinches, house finches, pine siskins and redpolls, are attracted. Where house finches are driving other birds away, look for feeders with perches above the holes. These will keep all but the most acrobatic birds from hogging the food. (45) Another variation has plastic inserts at the ports. Instead of a perch, each port offers a triangle that lies flat beneath the hole. The triangle is divided by narrow horizontal bars. In theory, only the feet of smaller birds will fit the openings between the bars. Another variation on the tube feeder is the Evenseed Silo (Hyde Bird Feeder Co., Waltham, Mass., 02254), a rectangular plastic box that splits in two for cleaning; perches double as bolts to hold the feeder together. The rectangular tube is divided into three sections, so that three types of seed can be fed at once.

Tube feeders with ports that are merely slits are meant for feeding niger thistle seeds and sunflower chips. Goldfinches will brighten these feeders, and redpolls, pine siskins, and occasionally indigo buntings will also frequent them. If house finches become a problem, shorten the perches with a hacksaw to 5/8 inch. (45)

Customers of one Midwestern dealer reported that the most successful thistle feeder he carried was a double tube feeder with four vertical plastic sticks arranged around each of the two tubes. The trick here was that only agile goldfinches could hold on long enough to get to the seed. Each tube had 20 openings. One customer was so impressed with the size of the flock of goldfinches he attracted that he gave the dealer a photograph. Unfortunately, the manufacturer stopped making that model.

In addition to feeders for seeds, there are several other specialty feeders available to build or buy.

Suet can be fed from a wire cage, or from a pine cone dipped in melted suet. Chicken wire with one inch square openings can be draped into a square or rectangle around a chunk of suet and nailed to a post or hung from a tree branch; the drawback to this inexpensive feeder is that birds' tongues might get stuck to the wire in cold weather. A mesh onion bag will serve as an inexpensive feeder without this danger. If you hang any of these feeders so that birds have to hang upside down to reach the suet, your feeder will repel starlings, whose weak leg muscles won't let them feed upside down the way chickadees and woodpeckers do.

Peanut butter will attract a variety of birds. A simple log feeder will hold this treat. Drill six to 12 holes, 1 to 1-1/4 inches in diameter and 1 inch deep, into a piece of log 14 to 18 inches long and 3 to 4 inches in diameter. Also sink a screw-eye into the end of the log so you can suspend it from a tree limb. The peanut butter gets smeared into the depressions. Even simpler, peanut butter can be smeared onto the bark of a tree. (45)

Various foods can be impaled on nails driven into a tree or post — apple slices to draw in gray catbirds, blue jays or robins, oranges for orioles, ears of corn for squirrels.

Simplest of all, seed can simply be scattered onto a well-drained patch of ground for birds who wouldn't come near an elevated feeder if their lives depended on it. Juncos, mourning doves, thrashers, sparrows, cardinals and towhees are used to searching for their food on the ground. (111) Leave only enough seed for one or two days and move ground feeding areas often. (107) To control disease, be sure to rake up the seed and compost it once a week, especially if it has been rained on. Another option is to spread the seed next to low ground cover like vinca or ivy, where birds can seek shelter. (111)

For more information on building your own bird feeders, Carrol Henderson's *Wild About Birds* is an excellent reference. It's available from Minnesota's Bookstore, 117 University Avenue, St. Paul, Minn., 55155.

BIRD BILLS AND BIRDFOOD

What a bird eats can be figured out by its bill. Warblers, creepers and other birds that skim insects from the surface of tree leaves or bark usually have slender bills. If seeds are the food of choice, birds like sparrows, buntings and other finches will have short, thick bills that they can use for cracking and husking seeds. Birds like crows who eat whatever they can find will have bills that are somewhat slender but also somewhat thick. (35)

A woodpecker's bill is shaped like a pick ax with an end like a chisel. Behind that bill, an arrangement of bone and muscle, ending in a short tongue, is stored by being coiled around the skull beneath the skin. Actually, the woodpecker's entire body aids it in eating. Strong, grasping feet and stiff tail feathers combine to give it a stable triangular base that supports it while it pecks in its search for insects. Once food is found, the long tongue structure is deployed to reach and surround the insect beneath the bark of the tree. (35)

SUNFLOWER SEEDS

According to Carrol Henderson's book *Wild About Birds*, any birdseed mix should contain at least 75% black oil sunflower seeds. (45) However, the birds that prefer sunflower seeds may sort through the mix and toss out the other seeds. If you don't have any ground-feeding birds who are willing to settle for hand-me-downs, this selective feeding will cause a lot of waste. A solution is to provide a separate feeder with nothing but sunflower seeds. (34) Keep track. If nobody is coming to the feeder with the mix, give up and feed only the big, black, oily seeds. Though blue jays, pine grosbeaks, tufted titmice and hairy and red-headed woodpeckers prefer the striped sunflower seeds that also show up as human snacks, black oil sunflower seeds are more acceptable to small birds like nuthatches, chickadees and finches because their edible hearts are easier for the birds to get to.

Sunflower seeds are such a hit with so many birds that you will probably find it more economical to buy them in big quantities. (107) But different birds attack the problem of eating them in different ways. Cardinals, grosbeaks and larger finches squeeze the seeds with their beaks until the shells pop open, but a goldfinch will have to find just the right angle before the seed will pop open for him. Chickadees hold the seeds between their feet and pound them with their beaks until they make a hole through which they eat whatever they can get at. A tufted titmouse or a nuthatch will find something to hold the seed for it, like a crevice in tree bark or in the wood siding of your house. Birds like doves and pheasants swallow sunflower seeds whole. The shells are softened in their crops, then ground in their gizzards. (34)

A blue jay will take away as many seeds as it can carry in its bill. When it finds a safe spot, it spits out all of the seeds, then eats one at a time. Whatever it can't finish will be left, and eventually may be eaten by another bird who didn't come to the feeder. Glenn Dudderar, writing in *Nature From Your Back Door*, suggests that people who don't like to see piggish jays at their feeders should remember that the seeds will probably get used by somebody somewhere.

OTHER FOODS

White proso millet spread on the ground will attract indigo buntings and mourning doves in summer, white-throated sparrows, juncos, white-crowned sparrows and fox sparrows in spring and fall. During the winter, juncos and American tree sparrows will eat it. Be aware, though, that this feed will also attract house sparrows, which will sometimes compete with songbirds for nesting areas, and brown-headed cowbirds, which lay their eggs in songbirds' nests.

Many species will also respond to cracked corn scattered on the ground, but this feed also attracts cowbirds and house sparrows. Milo sorghum will also attract many types of ground-feeding birds, and house sparrows don't care for it. (45)

If you're feeding a mix that contains a lot of proso millet and some cracked corn and you notice that you're attracting a lot of aggressive grackles, cowbirds and starlings, switch to sunflower seeds or to a better quality mix. (111)

Niger thistle seed is the preferred seed of goldfinches, redpolls and pine siskins. As winter begins, male and female goldfinches seem equally drab. But as the season progresses, yellow gradually creeps down the males' breasts. Then, whenever it seems that winter will never end, a glance at those expanding patches of yellow is a reassurance that spring is just around the corner, no matter how hard the snow might be falling. Niger thistle seeds are cheaper than anti-depressants, and they have no side effects. Unlike standard thistles, these seeds won't even sprout.

In winter, raw suet provides woodpeckers, nuthatches, catbirds, mockingbirds, chickadees, titmice, wrens, shrikes, thrushes and warblers and brown creepers, among other insect-eaters, with the protein they'd get from insects at easier times of the year. (35) If you decide to feed suet, you should hang the feeder on the north side of a tree or post so that the fat will spoil less quickly. Discard suet in warm weather, because melting grease will coat a bird's feathers, decreasing the feathers' ability to insulate. Grease on feathers can also lead to skin infections and feather loss. (107)

If for some reason you'd rather save the lives of the insects on your trees and your neighbors' by continuing to feed suet in warm weather, use rendered suet cakes that you buy in the store. (107)

If you'd rather feed your stale bread or cake to the birds than waste it, the birds will respond appreciatively. Bakery goods may not provide the best balanced nutrition, but they do draw birds to a feeder. Then it's up to the owner of the feeder to also provide what the birds actually need. (49)

Table 4-4 will show you who eats what.

SQUIRRELS AND FEEDERS

If squirrels keep robbing your pole feeder, you can add a metal guard at least 18 inches wide and four feet off the ground. If trees or wires are so close that squirrels can jump from them onto your feeder, put metal or plastic "umbrellas" above the feeder. Some feeders already come equipped with anti-squirrel screens surrounding the feeder tubes. You can retrofit an unscreened feeder with a wire-mesh cage. Use wire that the smallest birds can slip through and

A chickadee investigates Glenn Donovan's squirrel-proof feeder.

A closeup of the feeder shows the horn bells that hold black oil sunflower seeds. The lowest circle is one of the openings that had to be retrofitted to keep the squirrels out.

larger birds can reach through but squirrels can't maneuver through. (107)

Glenn Donovan, a Wisconsin artist whose medium is recycled materials, decided to create a squirrel-proof bird feeder using transmission gears and a junked spool of stainless steel wire. Wire mesh on the bottom of the ball catches stray black oil sunflower seeds that spill out of their holder — four recycled brass cornet horn bells.

When the feeder was first installed, fox squirrels quickly discovered that a few of the openings in the gears were big enough to slip through, so Donovan retrofitted by welding slightly smaller gears inside, decreasing the diameter of the openings by about half an inch. The morning that the squirrels discovered the addition, they ran around and around the outside of the ball, chattering angrily at their former chef; nevertheless, they were unable to join the chickadees, grosbeaks, juncos, cardinals, nuthatches and finches that flock to the feeder. To clear out seeds that have grown moldy, Donovan simply rolls the feeder down a hill.

A commercially available Gilbertson PVC feeder has perches made of spring steel. If the feeder is hung at least five feet from a tree trunk on a wire at least three feet long, birds will be able to fly the perches with ease, but squirrels will develop so much momentum that they won't be able to remain on the perches long enough to eat. (45)

Of course, you could break down and give the squirrels something to eat, like ear corn impaled on nails pounded into a wooden tray or post or tree, or nuts put out where the squirrels can find them. (34)

FEEDER MAINTAINENCE

If you're not feeding all year long, when your bird feeders come out of storage in fall they should be thoroughly cleaned before they are used. If there are brown creepers in your neighborhood, they might do some of the job for you. These insect-eaters will arrive to take care of spiders and other leftover insects. When the cleanup is through, the creepers will disappear. (34)

Steps can and should be taken to keep down the spread of disease at bird feeders. To counteract aspergillosis, a fungal disease that causes breathing problems, watch out for moldy food. Check your birdseed regularly, especially if it has been in the feeder for a long time. Check the feeder after a rainstorm or snowstorm and replace any wet food. (107)

Prevent salmonella poisoning and other diseases by cleaning up bird droppings periodically, and by disinfecting emptied feeders once or twice a month by soaking them for two or three minutes in a solution of one part chlorine bleach to nine parts warm water. For this job, vinegar is not a substitute. Rinse the feeders, and air dry them before you fill them again. (24)

If you notice sick birds at your feeder, stop feeding. Remove all the food, clean the feeders thoroughly, and wait at least 10 days before starting over. (107)

HUMMINGBIRDS

Because of the way hummingbirds feed and the food on which they feed, feeders for these birds are an entirely different breed. They come in two basic varieties. One is a ball-shaped (fruit-shaped in the cuter versions) or cylinder-shaped jar for storing sugar water. The water is fed into a basin and the birds feed through ports on the sides of the basin. On some models, a screen covers each port to discourage insects. Perches are optional or nonexistent.

The other type of feeder dispenses with the jar and simply provides a shallow bowl. The ports are on the lid and a continuous perch circles the bowl. This construction is wasp-proof because insects can't reach the low syrup level.

Both types should have one thing in common: red, the bull's-eye signal to attract hummingbirds. If for some reason your feeder lacks red, a red ribbon tied on top will do the job.

Like other bird feeders, though, a hummingbird feeder must be kept clean. Hummingbirds! Website evaluates several models of both types of feeders. It calls the bottle types "usually drippy and hard to clean," because a bottle brush is usually needed to do a thorough job. The bowl-types are "recommended for ease of cleaning and freedom from wasps and bees."

The feeder recommended most highly is the HummZinger, a clear, colorless basin with a red top that can be hung or set on a pole. It's from Aspects, Inc., 245 Child St., Warren, RI 02885, and is rated best for durability and ease of cleaning. An additional benefit is that the clear basin allows you to watch hummingbird tongues at work. Its durable continuous perch allows hummingbirds to perch comfortably while feeding, if this is their feeding style. (22)

When hummingbirds are hovering, their wings beat more than 20 times per second, sometimes as much as 80 beats per second, in a figure-8 pattern that allows the birds to take nectar from flowers that they couldn't reach otherwise. A hovering hummingbird uses eight times as much energy as one that is resting. (35) Though hummingbirds will flit from flower to flower, they will sometimes remain at one feeder until they are completely satisfied, so providing a perch saves precious energy for the little birds (22), if they choose to use it; some birds will ignore the perch completely and jump up and down like avian pogo sticks while repeatedly dipping into the sugar water.

A ruby-throated hummingbird zeroes in on a HummZinger feeder at the authors' living room window.

WHAT TO FEED/WHAT TO FORGET

As for what to use for filling the feeders, let's return for a minute to what hummingbirds eat: sugars from flower nectar for energy and insects for protein, vitamins, minerals and fats. (107) Hummingbirds don't return north until the insects return in spring, and there are usually plenty of those available naturally unless you've used insecticide, so the energy food is all that you can provide.

The standard mixture for filling hummingbird feeders is one part plain white sugar to four parts water, which is about the sucrose content (21 percent) of the flowers that hummingbirds favor. (22) Boil the water, turn off the heat, then pour in the sugar and stir until it's dissolved. Cool the liquid before you put it in your feeder, and store any extra sugar water in your refrigerator. Keep the refrigerated liquid no longer than a week. Don't substitute honey for sugar, because honey supports the growth of bacteria that can kill hummingbirds. (37)

A hummingbird can take in liquid equal to about twice its body weight in a day, and it weighs less than an ounce. Particularly in the spring, if one territorial male is driving everyone else away, you might want to cut down on the amount of sugar water you put into your feeder. (114)

There are a number of colored hummingbird foods on the market, and some people who make their own sugar-water nectar color it with red food coloring. However, natural flower nectar is clear, not red. A flower's petals are colored, and these are what draw the hummingbird's attention. The bird doesn't inspect the flower's nectar for color. The jury is still out on whether the red coloring is actually harmful to hummingbirds, but at the very least, it seems that there is no need for it. Another seemingly useless addition is scent in the nectar, for the hummingbird, like other birds, has virtually no sense of smell. (22, 107)

MAINTAINING A HUMMINGBIRD FEEDER

Before filling your feeder, clean it with plain, hot water or fill it with vinegar and uncooked rice and shake it vigorously. (107) If the solution in the feeder becomes cloudy or fermented, get rid of it immediately. Fermented sugar water causes enlarged livers in hummingbirds. In any case, empty the feeder, clean it and refill it every three days, or every two days if the temperature goes above 60 degrees. (37) Hanging the feeder in a shady spot will delay the onset of mold and fermentation.

The best way to prevent unwanted visits from ants, bees, wasps or yellowjackets is to avoid spilling sugar water on the outside of your feeder, or to wipe up spills. A dripless type of feeder also helps cut down on the unwanted visitors.

Yellow has recently been found to attract bees and wasps, so avoid feeders with yellow on them, like the yellow plastic "flowers" on the openings of some Perky Pet models. Either remove the yellow parts or paint them with red nail polish before setting the feeders out for the first time in the year. (22)

If ants start visiting your feeder, spread oil or petroleum jelly on the pole or chain that the feeder hangs from. Or dip a pipe cleaner in vegetable oil and wrap it around the wire of a hanging feeder. (107, 22)

If bees or wasps arrive despite your precautions, oil or petroleum jelly spread around the feeding ports should turn them away. (107) Another solution is to move the feeder a few feet. "Insects are not very smart, and will assume the food source is gone forever. They may never find it in its new location, while the hummers will barely notice that it was moved," advises the Hummingbirds! Website. If this doesn't work, remove the feeder for a day or two until the wasps stop looking for it. (22)

Because hummingbirds are so combative, providing feeders that have more than one perching area is not a bad idea, although pushing and shoving and outright harassing will probably continue as long as you've attracted more than one hummingbird at a time. (37) Every once in a while, a male becomes so aggressive that he keeps all the other hummingbirds away. This problem might be alleviated by buying a feeder with multiple perches or supplying several feeders. (107)

OTHER ADDITIONS TO A FEEDING STATION

In addition to birdseed and other edibles, birds will come for other things if you put them out. Many birds have gizzards that contain rough plates or ridges and take the place of teeth to grind up grains, acorns, nuts and other hard-shelled food that is swallowed whole, then rotated two or three times a minute. (35) Most of the year, the birds who seek it can find the grit their gizzards need to grind food, but when snow covers the natural sources, putting out parakeet or canary gravel might bring additional birds to your yard. (34) A small pile of sand or fine gravel in a space cleared of snow works well, too. (107) A gizzard extends a bird's menu by making additional foods available to it; this is an advantage, of course, unless the grinding aid is poisonous, like lead shotgun shot. (35)

Plant-eaters like finches need fairly coarse grinding aids, but fruit and nectar-eating birds and birds that eat soft-bodied insects may use fine sand or no grinding aids. On the other hand, there's nothing you can provide for a meat-eater like an owl, for its gizzard works by accumulating and compacting the inedible portions of its prey, then holding the result until the owl pukes out the pellet.

Ring-necked pheasant hens will look for calcium supplements in their grinding materials while they are laying eggs. (35) During nesting season, grit and crushed limestone or eggshells crushed into fragments smaller than the size of a fingernail will supply needed calcium so that pheasants and other birds can form stronger eggs. (107)

Crossbills, pine siskins and pine grosbeaks can use a little extra salt. If these birds frequent your feeder, provide salt blocks or mix rock salt with the grit or put it next to the grit pile. However, if deer are already a problem in your area, do yourself and your neighbors a favor and forget about supplying salt for the birds, because if you supply it, the deer will come. Also be aware that if you're feeding salt, it should be placed away from trees or other vegetation that you care about, because salt leaching into the ground could kill these plants. (107)

WATER AND OTHER CLEANING SUPPLIES

Besides providing food, you can provide another of the four necessities and a bit of *lagniappe* by buying the right supplies. Water is the necessity and a birdbath will provide a place where birds can bathe and preen. Like feeders, most birdbaths belong in clearings, where birds can see predators and have the extra time they'll need to fly for cover when their feathers are wet. However, some birds, including thrushes, like a ground-level spot in the

Birds should be able to see through or over plants as they bathe. (Sandra Stark)

middle of dense shrubs where they can hide and bathe. (37)

Robins, thrushes, mockingbirds, jays and titmice wade and force water to reach their skin by repeated motions. Swifts and swallows dip into the water as they fly. Chickadees, yellowthroats, wrens, buntings and water thrushes dart in and out of the water, immersing themselves and rolling briefly. Most woodpeckers and nuthatches and other birds with stubby, weak legs, bathe passively by exposing their feathers to drizzle. (35)

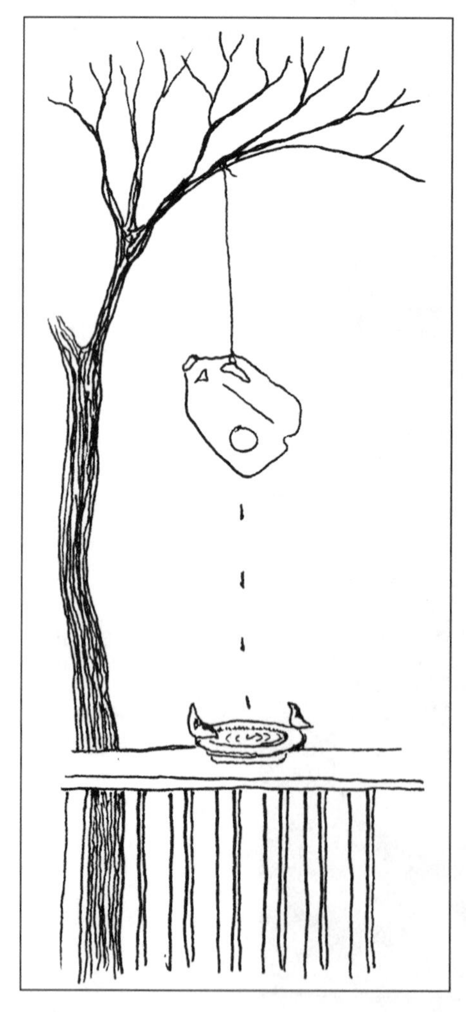

A drip arrangement suspended above a birdbath gives the illusion of moving water. (Patricia Gilbert)

In addition to sugar water, hummingbirds also appreciate water for bathing, but these hyperactive little birds prefer to take a shower. If you water your garden or grass, you might see them zooming over your sprinkler. Miniature sprinklers to attach to birdbaths are also available. (37, 4)

How often a bird bathes is weather-related. Titmice or chickadees might bathe five times during one hot summer day. In midwinter, they might bathe several times a week, often in protected areas where the snow has melted. (35)

Your birdbath should be shallow, from 1-1/2 to 3 inches deep. It should have a rough surface on the bottom and edges, preferably rough concrete, so that birds can walk out of the water. If you've bought a plastic birdbath with a smooth surface, put large, flat stones on the bottom to give birds a place to land that's above the water.

The splashing of water dripping into the birdbath will attract even more birds than still water. Make a pinhole in the bottom of a plastic jug, suspend it over the birdbath, and you've got yourself — and the birds — the illusion of moving water. (107)

You can create another very simple birdbath by digging a depression beneath a downspout, lining it with rock, concrete or plastic, and letting your ground-level birdbath fill with water every time it rains. You could even build a system of interconnected birdbaths to drain water away from your house. (67)

Heaters are available to keep your birdbath open all winter long.

As for the *lagniappe*, that little extra something, think about constructing a dusting area, a place where birds can wallow, shake the resulting dust over their feathers, and rid themselves of parasites. (43) Wrens, house sparrows, wrentits, larks, game birds and some raptors dust when standing water is not available. They will create their own dust wallows and throw the dust over their bodies, working it into their feathers, then shaking it out. (35)

Particularly if your neighborhood is a combination of sod lawns and black-topped roads and driveways, with little or no such loose soil available, providing a dust bath as a fringe benefit might bring you a bunch of new customers. Simply choose a 2-foot by 3-foot area in a sunny corner of your yard. Spade the rectangle to expose the soil, then keep the dirt fine, loose and free of vegetation until the birds take over the whole process for you. (43)

NEST BOXES

Where there are not enough dead trees to provide houses for cavity nesters, or where you'd simply like to be able to predict where a nesting bird is going to be, building a nest box is the ticket to possible success. If you have a woodlot that consists of nothing but pole-size trees that are too small for nests or dens, provide nest boxes for animals and birds that would nest there if the trees were older. (43)

By now you have probably guessed that what a bird is used to in its natural habitat is going to affect both construction and the location of the finished box. For instance, bluebirds' bills are too weak to rout out their own nest holes, but they will nest in old woodpecker holes. (4) So it's no surprise that they will readily accept a house that you build for them.

On the other hand, a bluebird will probably pay no attention to a house that is hung more than 15 feet above the ground, though a wood duck will not feel safe in a tree-mounted house that is lower than 15 feet above the ground.

For more particulars about the match between bird and box, see table 4-5 on page 225

NEST BOX CONSTRUCTION

Though the size of a nest box, the location of the portal, and its location will differ from bird to bird, some things remain true for all nest boxes. For starters, it must have a side or top that can be opened so that one year's

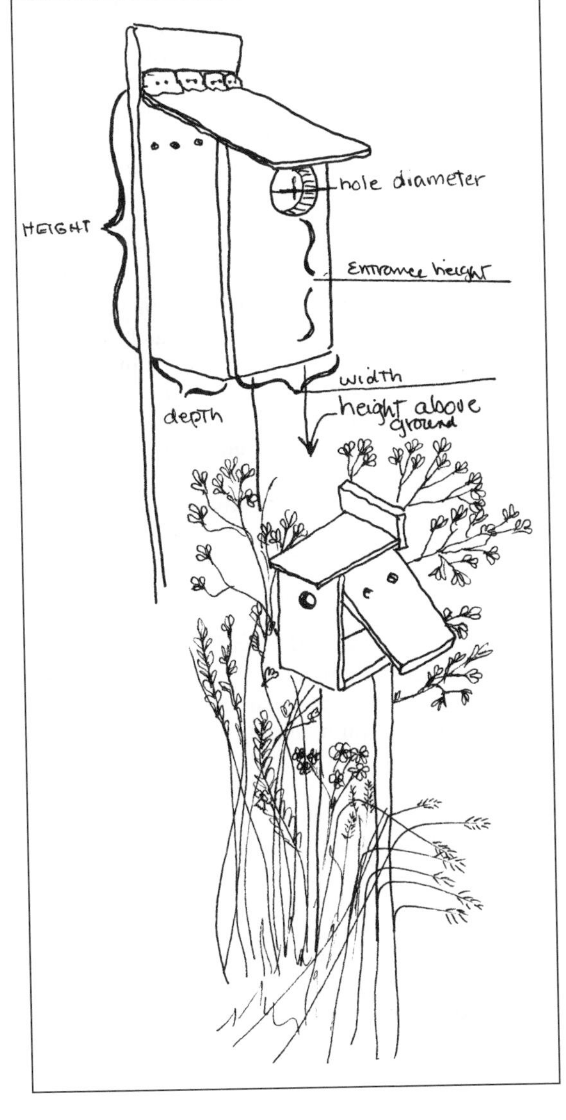

Decide which bird you want to invite to nest on your property, then use the dimensions in Table 4-5 to make this nest box the proper size. (Patricia Gilbert)

nesting material can be cleaned out before the next year's birds move in; the hinges on the door should be rustproof.

At least four 1/4-inch diameter holes should be drilled in the floor for drainage, unless you're building a triangular house with sloping floor and sides. There should be at least two one-quarter inch diameter ventilation holes near the top of the box's sides. The entrance hole should be just big enough for the desired bird to enter. A hole larger than 1-3/8 inches will allow house sparrows to enter; a hole bigger than 1-1/2 inches will admit starlings.

The nest box should never have a perch. House sparrows and starlings are the only birds that use them, and these are not birds that need help from you. If you've already bought a nest box with a perch, saw it off.

The front edge of the top of the nest box should extend at least two inches over the front wall for protection against rain and to keep cats that get on the roof from snagging birds as they emerge.

The sides of the box should cover the edges of the floorboard to keep rainwater from entering through horizontal cracks. The floorboard should be recessed a quarter inch from the bottom edges of the sides to prevent deterioration from moisture. (49)

Cans or other metal containers shouldn't be used because they heat up too fast; the exception is commercially produced aluminum martin houses.

Use 3/4-inch wood if you're building your own nest boxes. Pine will work for smaller boxes, but cedar, redwood or cypress should be used for larger boxes. Don't use creosote or wood treated with green, copper-based preservative. Large wooden duck boxes or other large boxes can be painted with wood preservative, but on the outside only. (49)

New nest boxes should be installed in fall so they can weather by spring. If you're fastening the box to a tree that leans, put it on the underside for better protection from the rain. (29)

If you have several different habitats on your property, experiment by putting a nest box or two in each. Place one or more in an edge environment, one or more in the middle of trees, and one or more on fence posts in the middle of a field or clearing. (43) To check on the results of your experiment, raise the hinged side of the box several times from the beginning of May to mid-June. Compare who comes to the boxes and the outcome of the nesting. (43)

Almost all birds are territorial, so boxes should be at least 25 feet apart, or 300 feet apart for bluebirds. Purple martins are the exceptions. They live in colonies in the wild, and prefer multi-dwelling martin houses. (112)

Bluebird houses have become popular additions to wildlife havens, perhaps because a bluebird house mounted on a fence post in a field, or even hung on a telephone or electric pole will almost surely be occupied. Moreover, the mother bluebird seems to accept human visitors who periodically lift the wooden side of her nest to check on the progress of her brood. Because the bluebird prefers sparse cover around its home, choose a spot with poor soil to erect your bluebird house. (4)

If tree swallows move into one of your bluebird boxes, locate a second box beside it. The tree swallow will defend its box against other tree swallows, leaving the second box for bluebirds. (4)

CLEANING HOUSE

Even though you've personalized the nest box to the extent of making it fit the species you're hoping to attract, the bird will personalize it even further by adding grass or other stuff that's close at hand. Clean out any old nesting material, because parasites still living in last year's leftovers could harm next year's young. Also, if the box is clean in fall and full again in early spring, you'll know that somebody has been wintering over there. (111) Better yet, leave the door open after nesting season to keep mice from moving in for the winter, then defending their territory when the songbirds you were hoping to attract come house-hunting. (43)

If paper wasps move into a nest box, they might fill the box with their nest. If this happens, wait until a morning cold enough to numb the wasps, then sweep out the nest with a handful of grass or a stick. (87)

If you don't get around to cleaning the boxes before winter comes, be sure to empty boxes by the end of March, or well before nesting season begins in your area. (43)

For a special treat, you can provide nesting material. Hang up a wire suet feeder or small wire basket filled with old dog or cat hair, short strips of cotton cloth, excelsior (shredded paper) or string or yarn in lengths of 8 inches or less. Be sure that any material you provide is biodegradeable. The next time you clean out your nest boxes you'll discover just where at least some of that material went. (37)

BRUSH PILES FOR WILDLIFE

One more structure that you might want to provide is a brush pile where songbirds, skunks, raccoons, opossums, woodchucks, chipmunks and rabbits can hide, keep warm, or simply loaf. On a larger property, quail, grouse and turkeys might join these others. For smaller, less mobile animals, brush piles should be located 100 to 200 feet apart. They should be within 150 to 450 feet of food or other cover and, ideally, within about a quarter mile of water. (89)

If you're not interested in a long-term brush pile, then simply gathering tree limbs and branches into one place will do the job for you. However, for more permanent installation, the base of your brush pile must be built of materials that are less likely to rot: stones, logs, or a combination of those.

The base of a pole brush pile should consist of four poles or logs 6 feet long and 4 to 8 inches in diameter, 8 to 12

A brush pile built to last begins with materials that won't rot easily, like the logs at the bottom of this pile. (Patricia Gilbert)

inches apart and parallel. Four more poles are then laid perpendicularly across the first four. Or lay three mounds of stones 8 to 12 inches in diameter separated by a Y-shaped space. Or combine the styles by starting with a base of four parallel logs topped with large flat rocks that span two logs. If these materials aren't available, lay one end of several poles on a stump to create a raised space in the center of the pile.

Once additional brush is stacked on top of these bases, larger tree limbs toward the bottom, the effect is to create sheltered tunnels beneath a 6- to 8-foot high brush pile about 6 or 8 feet in diameter. (28)

If you have a cluster of four to six saplings or small deciduous trees at least 4 inches in diameter at breast height, you could also create a living brush pile. Cut each tree only part-way through at a height of 6 inches to 3 feet, depending on the type of tree, the animal you want to shelter and the amount of access it needs at ground level. Leave a hinge of wood and bark, then bend each tree toward the center of the cluster. Done properly, a living brush pile will survive for three or four years. Create a living brush pile in the spring, after the leaves are out and the sap has risen. When the cut is made properly, the trunk should lie parallel to the ground. (89)

An alternative to creating a brush pile is to let the grass grow longer in ditches and along the edges of fields or lawns. Encourage the growth of woody cover (shrubs) in hedgerows and around ponds. (124)

These shelters, though beneficial to wildlife, may not be welcome additions near you or your neighbor's vegetable garden. Also, it's easy to mistake a properly constructed brush pile for a not-very-attractive dump. Before building one near your lot line, you might want to consult with your neighbors.

You might also look again at the list of animals a brush pile will attract. Being kind to skunks and raccoons might not be on your list of things to do.

There is one more thing to consider before deciding to encourage small mammals by creating shelter for them: far more than birds, small mammals are likely to carry disease and be a nuisance or danger to pets and humans. Still, if you like the idea of luring animals, keep your pets' vaccinations current, stay away from wild animals that are acting suspiciously and keep track of reports of diseases in your area. (107)

SQUIRRELS

If more gray squirrels is what you're after, check the trees on your property for food and cover opportunities. You're looking for mature trees bearing acorns and nuts. Red and black oaks will provide more food and a more consistent supply of it, but white oaks will be more likely to become den trees. Maples, elms, hickories, beeches, basswood, cottonwood, cherries and thornapples produce fruit or nuts that squirrels like. Plant new trees or encourage growing trees by releasing them — removing other trees that are shading them. (125, 32) Releasing or planting trees besides oak will give you insurance against a failure in the acorn crop.

What you're aiming for is a woodland that produces 150 pounds of acorns and other nuts per acre. This will maintain a good population of gray squirrels while making mast available to other species as well. You're aiming for mast trees at least 15 inches in diameter. (27)

Red squirrels are found in pine forests throughout the northern United States, wherever there are enough evergreens to supply their favorite food. They'll also move onto privately owned land, given the right kind of food. Pines are especially attractive, but red squirrels will also dine on other evergreens. (17)

In addition to brush piles and other low cover, you should plan on preserving snags and den trees for squirrels. An ideal nest cavity is 1 to 3 feet in length, and 6 to 10 inches in diameter, with an entrance hole 3 to 4 inches wide. Larger holes will admit raccoons. (27)

Where snags are missing in a forest that is too young, build and hang squirrel boxes. (125)

If you've decided to give in and accommodate the squirrels that appear at your feeder, be aware of a study in which squirrels that were fed on peanuts lost weight, then died. Apparently they didn't get as much nutrition as they would have from their favorite natural foods — hickory nuts and acorns. (17)

RABBITS AND HARES

The Penn State publication, *Woodlands and Wildlife*, suggests that if you want to attract rabbits to a five-acre property, the ideal set-up is four or five 1/4-acre patches of dense brush, briars, evergreens or vines for escape cover, four or five quarter-acre patches of planted or native grasses, unmowed, where rabbits can nest, and several patches of tall perennial weeds for late fall and winter food and cover. If there are no fields with food available, establish several long, narrow food patches. (43)

Charles Cadieux, author of *Wildlife Management on Your Land*, takes a slightly different approach: perfect rabbit habitat, he says, consists of 10-foot wide stretches of good cover between 10-foot wide food patches, and every 100 yards, a brush pile.

Snowshoe hares, found in northern areas of northern states, dine on low grasses and herbaceous plants in summer. In winter, they feed on woody twigs, bark and evergreen needles. They prefer their woody browse less than 6 feet tall, eating aspen, maple, birch, viburnum, spruce and white cedar. For winter cover, they must have dense patches of evergreens with branches that reach the ground in 10 to 20-year-old stands of spruce/fir forest. If you're looking for a project to attract or keep these animals, clear-cut your woodlot in blocks. (106)

Chapter 5

FINDING YOUR NEW VISITORS

So, now you've invited wildlife to your property, and you have reason to believe that they have accepted. Maybe you've seen deer tracks in the winter snow, maybe you've heard a barred owl hoot on a February night, or a pileated woodpecker hammer away on a sunny afternoon, or a turkey gobble at dawn.

Wouldn't it be nice to see these wild citizens who now have made your property part of their property?

It would indeed. But there are difficulties.

The basic difficulty is that it is the business of most wild citizens not be seen. A predator's life would soon end by starvation if it were constantly visible. And its prey's life would be even shorter if it were constantly seen. To have dinner or to be dinner, that is the question.

Most wild citizens would especially not like to be seen by humans. We are recognized by all, rightly or wrongly, as the superpredator, someone it is wise to stay away from.

And, remember, those animals are smarter in the woods and fields than you will ever be. It's their home, and they know far better how to hide than you know about finding them.

Having said that, there are ways of sneaking about that help even the odds somewhat. Many of those ways have been discovered by human hunters, who for thousands of years made their living by getting under the radar of the critters out there. Those hunters were — and still are — the superpredators, and those of us not of the hunting persuasion can learn much from our brothers and sisters.

Although a fisherman from very young boyhood, I didn't take up the other blood sport until I was more than 40 years old. I had much to learn, and fortunately I had friends who were excellent teachers. Here are some of the things I learned:

If you want to see wildlife, go alone. Two people make 15 times as much commotion as one, mostly because we insist on communicating, through talking, whispering, gesturing, whatever. If you decide to take a kid on a nature hike, that's wonderful, and I've done it many times, but try to impress on the kid the necessity to be quiet.

You might see more animals if you are in a car or pickup truck, because animals don't seem to understand that vehicles contain people. That's why you can drive by four deer feeding in a field near the road and they won't raise their heads or flick their tails, but you could never walk that near those same deer. You probably wouldn't see them at all, the deer having sneaked away before you were aware of them, or, if you did surprise them, you might see four white tails bobbing away in the distance. If you can drive a car to the edge of your wildlife habitat, and perhaps take a book along to keep you quietly occupied, you might see more animals than if you walked in and sat on a stump.

All animals' eyes and brains, including yours, are programmed to notice something in motion. That's what makes driving a car possible for us; we take note of that semi truck coming toward us, while ignoring the telephone poles along the same road. So if you must walk, do so slowly and carefully, one step at a time, and then find someplace to sit comfortably for a long time. Maybe on a stump, maybe on a folding stool. And then sit quietly, remaining as motionless as you can.

Animals may or may not be color-blind. There is much disagreement among scientists on that question. But there is no question that animals notice bright or white colors, especially in large sizes, and that they notice patterns of differing colors. Hunters — and soldiers — get around that problem by buying dark clothes that that have random patterns. The cheapest such clothes are found in military surplus stores, the best are found in stores that cater to hunters. Don't forget a camo hat, and maybe even a face veil.

Find natural vegetation to blend into. Think of a duck hunter's blind. Use a patch of brush as a blind, or create a blind by using tree branches and such. For a blind at the edge of a hayfield, it would be hard to beat one made of hay bales. Bear in mind that it might take a little time for animals to accept the blind as part of their world. If you want to watch migrating ducks, it's far better to build a blind in August than in October.

Think of using decoys. Duck hunters do, with great effectiveness, as do turkey hunters and even some deer hunters. A comprehensive outdoor-equipment catalog, such as Cabela's, *has decoys of both game and non-game animals of a surprising number of species. Decoys work in two ways: They attract animals that are interested in fighting or having sex with other animals of their own species, or, especially in the case of ducks and geese, they make critters feel they are in a safe place. In the latter case, a couple of heron decoys might attract ducks, and perhaps vice versa.*

Calling devices are used by duck hunters, turkey hunters and some deer hunters. But it takes special knowledge and skill to be able to sound an attracting call rather than an alarm call. I've been told that a device that squeals like a dying rabbit will attract predators. I have no firsthand experience with such a call, nor do I know anyone who has used one. It might be worth a try.

Each animal has an invisible circle around it, with the size of the circle depending on the species and circumstances. If a human steps into that circle, the animal leaves. That's what makes bowhunting so very much harder than hunting with a rifle. But where a hunter can sometimes stand outside a deer's circle with a rifle, a non-hunter can sometimes do the same with a good pair of binoculars or even a spotting telescope. And, of course, a camera with a telephoto lens.

Many animals are only or mostly active at night. There is much to be said for sitting on a stump in the dark night, listening to the busy life around you. Or sitting on the back porch and listening to whippoorwills and owls. Or, as many people are discovering, buying a camera that is tied to a tree or post and whose shutter is snapped only when an animal breaks a string or an invisible energy beam.

Perhaps the best way to see birds is simply to hang a feeder in the yard. Birds don't seem to notice people, or even cats, in a house, so you can drink your morning coffee in your warm living room while watching goldfinches and grackles just outside the window.

That works for me in the dead of winter.

— Ron Leys

MORE OPTIONS FOR MORE SPACE

The more land you own, the more different kinds of habitat you can provide for wildlife. You also have room to maintain each type of habitat in a variety of stages. And since each wildlife species has different needs, this means that you can expect a larger variety of animals, birds and insects as payback for your efforts.

Once you've discovered who is already visiting your property and what attractions your land already holds for wildlife, you can go in two different directions. You can choose a species and work on aspects of your landscape that appeal to that species, or you can choose one of the elements that is lacking and work on improving it. This chapter will tackle the problem both ways.

Ruffed Grouse

It's spring, mating season. Wherever ruffed grouse live, males are slipping easily through the dense vertical cover formed by young aspen or hazelnut or prickly ash or willow or dogwood shrubs to find stages 10 to 14 inches above the ground. A stage can be a log or rock or the root hummock formed by a downed tree, anywhere that a grouse can stand and beat his wings against the air in order to make the characteristic sound that earns him the name "drummer." That sound is meant to be conspicuous, to alert any female grouse in a fairly wide area that Mr. Wonderful has arrived.

To the human ear, a drumming male grouse might sound like an old John Deere tractor, but to any predator lurking in the area, the sound signals that dinner is at hand. Which is exactly why the most important aspect of the spot the grouse chooses is the amount of vertical cover it has. Researchers have found that a male grouse will look for an area with at least 2,000 dense vertical stems more than 5 feet tall per acre, or 60 to 170 stems or trunks of 8- to 30-foot-high trees within 10 to 12 feet of the drumming log. The ideal drumming territory will also be within sight of flower-producing male aspen, know in some areas as popple. (32, 42, 131)

Horizontal cover won't serve as well, except in small patches, because cover like slash — tree branches and other logging leftovers — tipped or blown down trees, or other forest debris could keep a grouse from seeing predators or moving freely to avoid them. (65) When a bird of prey spots it, a grouse is most likely to escape if it can fly through cover where its short wings can pass but a large hawk or owl can't pursue it. (42)

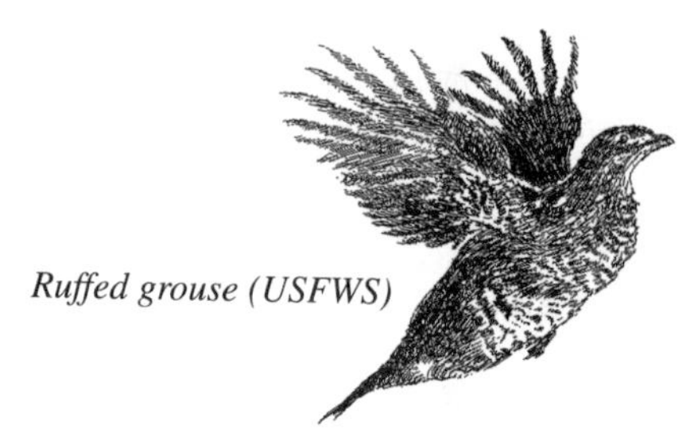

Ruffed grouse (USFWS)

A male grouse will defend six to 20 acres of home territory. Although at least 40 acres is usually needed to manage for this popular game bird, 20 acres of aspen, oak and shrubs, plus a neighbor's woods, might support one or two drummers in spring, perhaps a brood of new chicks in summer and a few grouse in fall. But what borders your acreage is also important. Even 50 acres of perfect habitat might not attract grouse if it is completely surrounded by open fields with no safe travel lanes connecting to it. (32)

ASPEN

Ruffed grouse is the favorite game bird in the United States, and aspen or popple is its favorite tree. The buds, twigs, catkins and leaves of aspen provide food, while the straight, toothpick-like trunks of young aspen trees give just the sort of vertical cover grouse need. (32) No other plant in all its life stages does as much for a wildlife species in whatever life stage it finds itself. (42)

If you're managing a stand of aspen to attract grouse, aim for three age classes, because grouse need three types of habitat within 40 acres to thrive. Aspen that is less than five years old will provide brood habitat for hens and young chicks and might possibly shelter a drumming male. Aspen six to 25 years

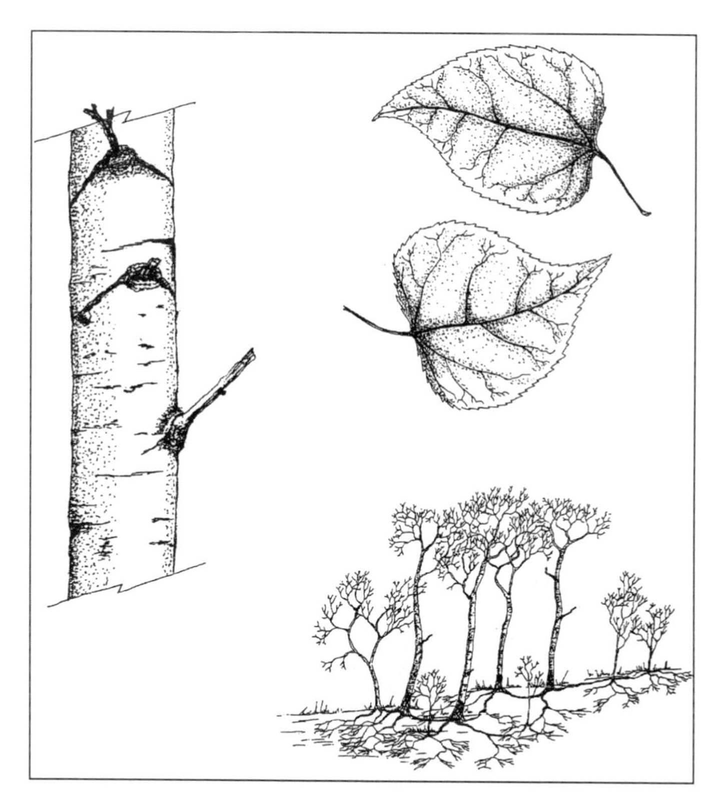

Aspen (Bureau of Endangered Resources, Wis. DNR)

old will give cover for drummers, nesting hens and wintering adults. Aspen older than 25 years will provide buds and catkins for winter food and nesting and brood-rearing cover if the right shrubs and non-woody food plants are also in the stand. (32)

In New England, where the sun-loving aspen is decreasing in forests, grouse numbers are dwindling, even when those forests contain birches, beeches and cherries, considered alternate food for grouse. In northern forests, where snow is on the ground two to four months a year, grouse live in moderate to high density only when aspen is present. (42)

Fortunately for anyone trying to attract grouse, aspen is very easy to clone, given the right circumstances and the right treatment. If you are planting a new patch, the poorer the soil, the better, so that trees in each section remain in a stage of succession as long as possible.

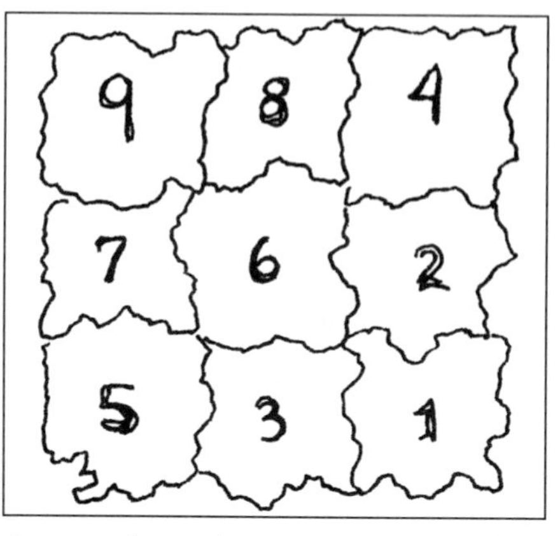

A pattern for cutting aspen to maintain the stages of succession most useful for grouse. (Patricia Gilbert)

If you're revamping an existing patch, the best time to cut aspen is when trees are dormant, in the winter months between leaf fall and sap run. This is the time when the most nutrients are stored in the roots and suckers have the best chance of sprouting. (42) To maintain all of the ages of aspen a grouse needs, cut blocks of aspen in four- to five-year intervals. The illustration shows the pattern to use. Notice that the cutting proceeds from south to north, so that the fuller canopies of the older growth don't shade the newer growth. (42) Because aspen needs full sunlight, though, it probably will not do well if a block is less than one acre . (43) Be aware, too, that if hunting is not controlled, a block not much bigger than an acre might not give the resident grouse a large enough area for cover. (42)

Gordon Gullion writes that in order to maintain blocks of aspen, you must be ruthless and clear all of the visible aspen from the block that you're cutting. If you don't, the auxin hormone in the smaller, inferior sprouts that you've left behind will suppress what might have been superior clones. (42) Though Carl Tubbs, Richard DeGraaf, Mariko Yamasaki and William Healy concede in *Guide to Wildlife Tree Management in New England Northern Hardwoods* that leaving mature aspen in a block will reduce root-suckering, they point out that aspen rots easily, so it quickly produces good den trees; perhaps it's worth sacrificing a few new aspen in order to leave den trees in the area, they argue. (110)

In the seasons when they are available, strawberries, blueberries, bunchberries and raspberries, parts of sedges, clovers, violets and grasses become important additions to the grouse's menu. In fall, berries of dogwood and viburnum, sumac, grapes and acorns supplement the aspen, while in winter, grouse will also eat the catkins and buds of alder, hazelnut, willow, beech, birch, maple and some berry bushes. (131) Grouse can survive harsh winters because their lower digestive tract can handle a coarse diet of buds and the green parts of plants, which is often the only winter food available to them. (68)

The few animals grouse eat include ants, beetles, flies, spiders and other such creatures. (131)

COVER FOR GROUSE

Grouse welcome snow, for when snow is soft and deep enough, they will burrow 7 inches or more beneath the surface and burrow-roost in the coldest winter weather. A picture of such a burrow shows what appear to be two footprints in deep snow. According to the caption, a grouse in full flight plunged into the snow, roosted overnight in an outside temperature of 20 degrees, then emerged in mid-afternoon after flushing from the burrow into full flight. In parts of the country where snow is an important factor, planted food like berries and skid roads and log storage landings seeded with white clover are less important to a grouse's winter survival than the amount of snow on the ground. (42, 131)

Evergreens with low-growing, bushy branches also make good winter cover where good roosting snow isn't available. For dry (xenic) sites, a Wisconsin publication recommends eastern red cedar, white spruce for medium to moist (mesic) sites and eastern arborvitae on wetter (hydric) sites. (32)

Balsam fir is one type of evergreen that provides winter cover for grouse, but this species will take over a stand of aspen, as will northern hardwoods. (32) Clumps of evergreens should take up no more than 30 percent of a stand of trees. The evergreens must retain their lower branches; otherwise, they make perfect perches for predators like hawks and owls because the remaining branches of the evergreens keep grouse from seeing the alarming silhouettes, when bare deciduous trees would expose them. (32)

CREATING ATTRACTIONS FOR GROUSE

For brood cover where little or no aspen is available, the edges of old fields, road edges and the edges of forest clearings may be useful. To create a forest edge, cut openings of up to one acre in the forest or cut down trees that grow in the swath 30 to 50 feet from the edge of the road, then allow brush and weeds to grow. Maintain this edge by periodically cutting trees. (106)

According to Gullion, author of *Managing Northern Forests for Wildlife*, grouse prefer good cover lacking a definite edge to a sharply defined edge, unless the only cover available is at the edge of a woods that is too mature. A Minnesota study showed that drummers will choose a drumming log as far from a definite edge as possible, given the choice. (42)

If you have aspen on your property but drumming logs are scarce, you can create one. Saw down a tree that's 10 to 14 inches in diameter, or pile two logs to get the desired height. Leave a 2- to 3-foot stump and keep the log close to the stump as a guard object. Your log should be in an area of good vertical cover, but in the circle within 10 to 12 feet of the drumming log there should be no logs or trees that can impede the grouse's vision. If there isn't already good vertical cover, clearcut all of the aspen in the vicinity and it will quickly grow back.

As for providing water, grouse are different from backyard songbirds, for they seem to get enough water from the food they eat and from dew that they find on that food. (32)

If there are no grouse on your land because one of the four essentials is lacking, will you have success if you provide what has been missing? According to Gullion, if there are grouse within two to three miles, and there is fairly continuous forest cover between existing grouse coverts and your land, young birds could find your property in the course of the normal fall dispersal. However, if they have to cross more than a quarter mile of water or half a mile of open fields, they won't, for they simply are not strong long-distance flyers. (42)

Bobwhite Quail

Bobwhite quail like to nest in open, grassy areas, with brush nearby for escape cover. They'll nest on the ground in fields, in grass under orchard trees and in forest openings. (43) The nests will usually be no more than 60 feet from the edge of a road, path, pond or forest, in grass that the hen and chicks can see through easily. Quail can't travel through matted ground cover. (17)

The brush must be dense but does not have to cover an extensive stretch of territory. (68) Quail will often use thickets of hazel, raspberry or grapevines, dogwood shrubs, willow and elderberry for cover. Many are found on small farms with hedgerows and scattered plantations of evergreens. In Wisconsin, the quail population declined when farms became bigger and more intensively farmed and brushy areas considered waste were removed. (130)

Quails' ideal range contains equal amounts of cultivated crops, ungrazed grass, brush and woodland with an open canopy. (43, 130) Adults eat fruits,

Bobwhite quail (USFWS)

greens, seeds, the fruit of wild grape vines, bittersweet and sumac, and, occasionally, the seeds of native legumes. Adults will also occasionally eat insects like grasshoppers, crickets and beetles. In late fall and winter they will eat large weed seeds of plants like ragweed, foxtail, smartweed and wild buckwheat. They'll also forage for waste corn, wheat and other grains, sassafras seed, shattered acorns and soybeans, and, in winter, dine on the corn they can find after a manure spreader passes. (130, 51)

Beautyberry, blackberry, dogwood, holly, gallberry, hawthorn, honeysuckle, roses, sassafras, sumac, viburnum, wild cherry and wild grape will provide both food and cover. Black locusts provide food but little cover. The only oaks that benefit quail are those that produce acorns small enough for the birds to eat whole. (17)

A food plot of corn, wheat, oats, buckwheat, sorghum, millet, rye and/or clover is an attraction for bobwhite quail, especially if it is close to winter cover. (43) A food plot may look like any other crop field, but the produce is only meant to be eaten by wildlife.

Water in several forms is crucial for quail. Quail need water to drink during part of the year. Also, when the humidity drops, quail eggs won't hatch because the egg membranes have to remain soft and pliable so that the babies can peck through, a process that can take much of a day. (17)

At the northern edge of the quails' range, cold won't hamper them if there is enough winter food that is tall enough to stand above the snow. Dense thickets for shelter in winter are crucial for quail. Privacy is, too, for if a bird is flushed in winter and snow hits the inner feathers under its wings, the snow will stick. When the bird lands and folds its wings, the snow melts and lowers its body temperature, sometimes far enough to kill the tiny bird. (17)

If food and cover are both available, quail will move no farther than a quarter mile a day, though their seasonal movements average about one mile. (43, 130) Where both food and cover are adequate, roaming cats and dogs, raccoons, skunks and snakes won't permanently affect the quail population. (51)

INCREASING QUAIL NUMBERS

If you own land that once harbored upland game birds, don't assume that you can simply buy birds and dump them back onto that land. The authors of *Woodlands and Wildlife* advise that pen-reared birds will not survive and thrive. (43) But more important, the original gamebirds might have disappeared because the habitat changed. So before you spend any money on chicks, take a long look at the land you want them to inhabit and ask yourself what is missing. (17)

Provide shrubby cover along fences, woodlots and roadsides for travel lanes. These will be most helpful if they are right next to croplands, woodlands and fallow grasslands. Plant scattered shrub thickets to supplement existing hedgerows. Young pine plantations will offer inviting roosting cover. Stop mowing along roadsides and keep woody plantations from taking over idle grasslands. (130) Plant some shrub lespedeza (lespeza bicolor), a densely growing plant that provides some winter cover and lots of seed. (72)

If you're creating a forest opening for bobwhite quail, you'll have to expose at least 60 percent of the forest floor to sunlight to encourage the food and cover plants quail need. (43)

In grasslands and prairies where trees are absent, Payne and Bryant suggest building a teepee of 10 to 20 fenceposts or railroad ties, wired at the top and spread open at the bottom. These will provide a cool place for quail in the summertime by blocking the sun's rays.

Do some or all of these things, then wait for the quail to return. If they don't, try to re-stock with wild-trapped birds that know how to avoid predators. (17)

Woodcock

Woodcock winter in the southern United States, but their breeding range is from southeast Manitoba to Texas and east to the Atlantic coast. (68)

Twenty five acres is a large enough parcel to manage for woodcock. They prefer thick, brushy woods and old logging roads bordered by alders or aspen, dogwood or willow swales for nesting and feeding. Cutting virgin forests improves woodcock habitat. (68, 13) Moist woods or brush pastures are the choices for nesting.

The best feeding for adults is alder swamps where they can find earthworms easily. They will also use wet or poorly drained brushy areas and young aspen and white birch stands, and will feed on insects in the leaf litter under the forest canopy or in tree bark. In summer, they stay in areas where there is heavy leaf litter but very little ground cover. One study in Pennsylvania found woodcock in both alder and aspen sites, where leaf litter is richer than it is in other types of forest. In Minnesota, when rainfall has been normal to heavy, woodcocks have been found to prefer aspen over all other trees, including alder, even though earthworms were available in equal amounts in aspen and alder areas and the leaf litter was equally rich. (42) A Vermont publication advises that many marshes in that state are too wet for woodcock. (106)

After the snow melt in spring, male woodcock return first, in Vermont as early as March. Females follow soon after. The males require openings, bare ground 15 to 20 feet in diameter, for their singing ground. These must be surrounded by shrubs or trees no more than 6 to 10 feet tall for a 10- to 30-yard radius so that the male can take off successfully at a low angle in order to display for the female attracted to his singing ground. (42) Courtship flights may take place in areas where blueberry, goldenrod, and red osier dogwood have begun to take over. (13)

Hens nest in forested sites or brush. Their nests consist of nothing more than scrapes in dead leaves. Newly hatched woodcock spend a few weeks in small brood areas that contain lots of earthworms. The chicks are independent in a month. (42)

In late summer and fall, woodcock roost in large, open, grassy fields or clearcuts with a lot of slash, where predators have less of a chance of finding them. (42)

After 20 years, alder and dogwood are too big to be useful for woodcocks, so a quarter of your forest should be clearcut every five years in 10- to 20-foot wide strips or quarter-acre patches. About 20 percent of the total area should be kept open for roosting and courtship rituals. Mow or run a brush hog over this section every three to five years. (13)

Wild Turkeys

The truth about wild turkeys is that they like to roam. Yearling hens in one study moved 8 to 10 miles from their winter range during spring dispersal. (71) They also prefer to stay in woodlots of at least 100 acres, although a wooded travel corridor that leads from someone else's woods may make your smaller area acceptable. Your county extension service should be able to provide you with aerial photos so that you can scout more than your own property. (70)

You won't be able to maintain a flock year-round on 40 acres. Instead, you must decide which stage(s) of the turkey's life cycle you want to encourage.

Wild turkey (Patricia Gilbert)

A clearing planted with grain crops that will stand above the snow in winter, plus clover or alfalfa will be an area good for brood rearing and summer foraging. If there are mast-producing shrubs in the opening, it will improve the food supply, but might decrease the acceptability for brood rearing. (70) If there's good nesting habitat, the gobblers might follow the hens to it. (32)

After breeding season, turkeys gather in small flocks. Young males go off to form their own flocks, while young hens join flocks of older hens. At this season, they will be looking for an area where the food they prefer is abundant enough to feed a number of birds at a time. (70) This often means that mature forests, especially those containing oaks, will be prime turkey habitat, because this is where the best mast trees are found. Also, mature forests usually lack thick understory, (6) a plus for turkeys, whose sight has been found to be 10 times as good as humans'. (70)

FOOD FOR TURKEYS

Many different foods satisfy wild turkeys, though 90 percent of adults' food is plant material. They will feed on the buds of trees or wander into bushes to eat berries. Acorns are their number one food choice, but because the acorn crop can be undependable, soft mast crops, like the fleshy fruits of cherries, crabapples, hawthorns and others should be favored if you're planting trees to attract turkeys. (40, 17)

In summer, adult turkeys will also eat tubers, snails, roots, flowers and fruits, but corn can make up half of their diets, with a quarter of their food coming from wild plants like dandelion flowers, weed seeds, wild grapes and gray dogwood fruit. Insects like dragonflies and grasshoppers make up about 4 percent of their diet. In summer, three-quarters of the turkey poults' diet is insects. In fall, the stomachs of birds shot by hunters show the amount of corn decreasing, the amount of wild plants and insects increasing slightly. In winter, turkeys will eat leftovers from fall — green plants still growing around seeps, nuts, seeds and fruit. (82) A sure sign that there are turkeys on your land is the characteristic scratched-up patch in the forest duff where the birds have been looking for acorns. (71, 82) If winters are very harsh, turkeys will sometimes even eat tree buds. (70)

In farm areas, waste grain and silage provide part of the turkeys' winter diet. The sight of wild turkeys following a manure spreader through a bare winter field is not uncommon. What the turkeys are after is not the manure itself, but any corn that the manure-producers were unable to digest.

And how can wildlife biologists be so sure that the corn turkeys consume is waste corn? They recognize it by its weathered appearance, the teeth marks of other animals on it, or the manure smell that lingers when a turkey's crop is opened. (71).

IMPROVING THE TURKEY FOOD SUPPLY

During winter storms, turkeys can spend a week or more on their roosts. Studies show that they can live without food for up to two weeks. However, if soft snow is deeper than 6 or 8 inches for more than five to six weeks, turkeys will probably starve. (82) On the other hand, in a year with little snow cover, it may be necessary to knock down standing corn so the turkeys can reach it. (17)

Oaks, hickories, cherries, beech, hop hornbeam, and ash provide mast. Turkeys also will eat juniper berries as well as the fruit of barberry, dogwoods, viburnums, hawthorns, grapes and other food-producing shrubs, and burdock and sensitive fern. If shrubs that turkeys favor are already growing on your land, but taller trees overshadow them, remove the overstory to favor these more valuable plants. (32, 106) You might also need to thin stands of young trees in order to encourage the most productive seed trees. (43)

To provide spring food for turkeys, unused roads and bare landing areas where loggers have stored logs for pickup can be seeded with cool weather grasses and legumes — an orchard grass and clover mix, for instance. Orchard grass greens up early and grows late if the weather isn't severe. Orchard grass will provide the turkey with protein in spring, and the clover will take over as a source of protein when grass production declines in summer. In addition, the clover will attract insects for the turkey poults. (40)

Areas where springs seep through the ground and the water keeps flowing during the coldest weather are magnets for turkeys. If you are lucky enough to own such a treasure, improve it by thinning trees to encourage the growth of green plants. The increased greenery will increase the supply of insects. While thinning, though, don't remove any seed- or berry-producing shrubs or trees. (43) Large mast-producing trees near seeps will drop their seeds, which will sprout and provide winter food. (40)

TURKEY "SPACE"

In western Wisconsin's Vernon County, the first area in the state where wild turkeys were re-introduced, DNR biologists found that in spring and fall, turkeys spend 62 percent of their time in woodlands, 23 percent in crop fields and 15 percent in pastures and idle woodlots. (71)

"Space" for a tom turkey includes a strutting zone that he will defend and where he can perform spring mating displays for hens who might be a long way off. The gobblers need areas where they are easily visible to the right animals, but close to cover where they can duck in if the wrong animals happen by. Open woods, grassy fields, trails, or the edges of hayfields provide these strutting zones. (70)

As well as food, turkeys need places where they can stay warm in winter. South-facing hillsides and areas with spring seeps are attractive because they have less snow. Evergreens and oaks that keep their leaves all winter can provide warm cover. These plantations should be arranged in clumps or belts along the edge of old fields or openings or in open woodlands. They should be planted close to the turkeys' winter feeding area. In Wisconsin, red and white pines are recommended, though spruce, fir and cedar will also provide acceptable cover. (70)

NESTING AND POULT-RAISING

At most stages of their lives, turkeys can usually make do with the moisture in their food and dew on grass and plants, (70) but during the nesting period, a hen needs a source of standing or fresh water within 200 yards of the nest, which is simply a grass-lined depression in the ground. (40) Good nesting territory will be a woodland with a good supply of grass and forbs like goldenrod, yellow coneflower and black-eyed susan. (70) A turkey hen lays 10 to 12 eggs over a two-week period. (82) The eggs hatch in 28 days, usually in late May or early June. If a turkey's first nest is destroyed, the hen will usually move to an area with more plant cover before she tries to establish a second or even a third nest. (70)

Turkey hens occasionally nest in alfalfa hayfields, where they are at great hazard of being killed by harvesting machinery.

After they hatch, the poults are moved to a grassy area to find the insects they need. (32) But the grass must not be so dense that it hampers the movement of the poults before they can fly (40), or the ability of the hen to watch for predators. When they are two or three weeks old, the poults will fly, and roost in trees at night, like their parents. (82) For roosting, turkeys prefer scattered trees with horizontal branches that are taller than the trees that surround them. Individual red and white pines and oaks are the trees of choice. (32, 70) In Vermont, hens and poults are often found in stands of hop hornbeam with grassy areas beneath. (6)

Poults need clearings with scattered trees where they can use the weedy cover while eating the protein-rich insects that live there. Clearings as small as a quarter of an acre can improve habitat for turkeys and other wildlife. (43) For cover, poults can also use a stand comprised of the stems of sapling or pole-sized hardwood trees for protection from birds of prey. In the grassy understory, with little or no brush, the poults have a clear view at ground level. (106) Agricultural land or other grassland that has not been sprayed, oak woodlots that have been burned periodically and other savannas are prime territory for turkey poults. Burning not only makes it easier for poults to see and escape predators, it also increases the insect supply. Hay and oat fields that have recently been cut are also attractive because insects do well in these areas, yet are easily seen and caught. (70)

INCREASING TURKEY NUMBERS

Perhaps the history of the return of the wild turkey to New York State is a cautionary tale for landowners who want to put more of these birds on their property.

At the dawn of white settlement, wild turkeys existed in all parts of New York State south of the Adirondacks, but by the mid-1840s, the birds had disappeared, over-hunted and driven out by landowners who cut down their forests for lumber, then transformed the land into small farms. Farming in the state waxed, then waned. The first land to revert to the wild was the most unproductive land at the tops of hills.

A century after the disappearance, wild turkeys from a remnant flock in northern Pennsylvania started trickling back over the border into southwest New York State. Encouraged by the presence of the new arrivals, wildlife biologists converted a state pheasant farm into a place for raising wild turkeys. But when those birds were turned out into the wild, they didn't have enough street-smarts to survive the neighborhood predators.

Meanwhile, however, in that southwest oasis, the genuinely wild turkeys were being fruitful and multiplying, either unaware of or unconcerned about the statistics that predicted that 60 to 70 percent of their progeny would perish before they reached adulthood.

After eight years, the professional bird-raisers stopped trying to re-create wild turkeys in pens and began to trap the successful wild turkeys and plant them all over the state. Today, according to a publication of the New York State Department of Environmental Conservation, the only places where you can't find wild turkeys are certain parts of New York City and, perhaps, neighboring Nassau County. (82)

Introduced wild turkeys have been wildly successful in a number of other areas. In 1973, about 2,000 birds were released by the Maryland Wildlife and Heritage Division of the Maryland Department of Natural Resources in a trap-and-transfer program. Today, there are an estimated 30,000 birds in that state. Because space is the survival element in shortest supply, the birds are found most often in protected forests along the rivers of central Maryland or in fields and woodlots in southern Maryland or on the Eastern Shore. (40)

The moral seems to be that you can work on turkey habitat, but be satisfied with the wild turkeys that come to your land naturally. Don't stock game-farm turkeys, for these could spread diseases to the truly wild birds. (32) Pen-reared birds lack a natural fear of humans; in fact, they might rely on humans to such an extent that during a hard winter, they might lack the natural ability to find substitute food sources on their own. (71) Pen-raised turkeys also don't have the upbringing or genetics to know how to respond to predators. When pen-reared birds mate with wild turkeys, the next generation might not be as well adapted to coping with the dangers of the wild. (70)

Pheasants

Pheasant (Patricia Gilbert)

Ring-necked pheasants, originally imported from Asia, thrive in rich farmlands, especially where grains like rice, wheat, and corn are grown. They need little in the way of tall cover, but do need escape areas such as fencerows, ditches, or patches of cattails or brushy woods. (68) If conditions are right, most pheasants won't move more than 3 miles from their birthplaces. However, if there are more than a few miles between dense stands of grass for nesting cover, feeding sites for poults and cattail marshes or shrub thickets for winter cover, the pheasant population might be so small that predators can easily wipe it out.

Hens make nests from April to early May, gathering grass and weeds into a nest bowl, a depression in the ground. Early nesters usually use last year's dried vegetation or wetland borders, hedgerows or roadsides. Later nesters usually use new grassy growth, like that found in hayfields. Most pheasants have hatched by the beginning of July. (127)

If you want to provide nesting cover, odd-shaped plots are better than long, thin strips along the edges of fields, since the thin strips are also used by meat-eaters as travel lanes. (17) A border at least 20 feet wide on each side of a stream or other water will give pheasants a good place to nest. (89)

Grazing and haying should be timed to avoid destroying pheasant nests, because mowing is the greatest source of pheasant mortality. Raising a hay mower to 5.5 inches (25 cm) when cutting alfalfa will save the lives of a lot of pheasants. But this is a hard thing to ask a farmer to do. In one study, raising the mower's cutting teeth to this height cost farmers $85 for every pheasant that was saved. (89)

PHEASANT COVER

In winter, pheasants need two types of cover: foraging cover to keep predators from seeing them while they're occupied with eating, and roosting cover that gives them shelter from wind and blowing snow. The pheasants' varied pattern of feathers usually provides adequate camouflage for most of the year, concealing the birds as they hunt for waste grain. In winter, however, that same pattern stands out against white snow, especially if that snow has buried the corn stalks that remained after the fall harvest. A pheasant might be able to scratch through the snow to find waste grain, but its cover is gone. Therefore, when foraging cover is buried, pheasants usually won't move more than half a mile from roosting cover, no matter how good the food supply might be elsewhere. If many pheasants live in an area, they will condense where roosting cover is good, lowering the amount of food available to each individual. (91)

Winter cover does not have to be complicated. Lowland brush, heavy stands of asters and lowland grasses, cattail or tamarack swamps will do. (54)

Ideal range for pheasants is flat or gently rolling land that is 70 percent cropland and 30 percent brush, marsh and woodland. (43, 127)

If you want to change your land in order to attract pheasants, keep on developing brushy cover in fencerows, gullies and along stream borders, and leave cattail marshes. You can also plant one to three acres of evergreens, or a full-fledged shelterbelt where pheasants can roost in winter. (43, 127) Weedy cover that stays standing tall even in snow will also give pheasants good winter cover, especially if it is bordered by a food plot. (17)

PHEASANT FOOD

Adult pheasants mostly forage on waste small grains and corn left in fields. They also eat weed seeds like burdock and ragweed, insects, plant leaves and minerals. During egg-laying, hens eat lots of snail shells and grit high in calcium. (127)

Young pheasants survive on the protein from insects. Again, a field that has been treated with herbicides, clean of all weeds, will produce far fewer insects (which is part of the point) and will therefore be less helpful in attracting pheasants.

Though pheasants can get by on what they find while foraging, you can help out by planting a food plot. Corn is the grain of choice for pheasants, for it will usually remain standing through repeated snowstorms. Corn planted in large patches provides adequate foraging cover, but planting grain sorghum with the corn can entice pheasants into feeding regularly in the area through several seasons. (91)

Food plots are especially important during hard winters, especially for pheasant hens who need to be in good condition during the mating season when, in a few short months, they will have to lay and hatch eggs, raise a brood, and molt. If hens are in poor physical condition during the spring, they may die. If their chicks are still young, there will be additional deaths.

More important than the food value of a food plot is its value as cover from winter storms. Wisconsin biologists have seen pheasants leaving an area with poor cover and good food for an area with poor food and good winter cover. (9)

For more information, read FOOD PLOTS later in this chapter.

INCREASING PHEASANT NUMBERS

As for stocking your land with pen-raised pheasants and hoping that this will permanently increase the number of pheasants on your land, the advice is: don't even think about it. According to information from Pheasants Forever, when land is stocked with eight- to 14-week-old pheasants, 60 percent may survive past the first week, but only 25 percent will be left after a month. After one winter, only five to 10 percent remain, with five percent being more normal. Predators will account for 90 percent of the deaths because these pen-raised birds don't seem to know how to avoid their enemies. Also, because they have been hand-fed, it may take pen-raised birds as long as three weeks to learn how to forage. (92)

Stocking with pen-raised hens in spring turns out to be no solution either. Before they even try to nest, 40 to 70 percent will die. Of every 100 hens remaining, only five to 40 chicks will be raised. Unfortunately, if you expect the pheasant population on your land to remain at a constant level, each hen should have four chicks that survive at least 10 weeks. (92)

There's more bad news: there is evidence that pen-raised pheasants might dilute the wiliness of wild stock if they do live long enough to reproduce. There's the additional danger that with an abundance of pheasants temporarily in the neighborhood, predators might develop more of a taste for them, turning to the truly wild birds once the supply of tamer birds has diminished. And, as always, there is a potential for disease to spread from pen-raised to wild birds. (92)

So how can you get more pheasants on your land? Not surprisingly, habitat is the key. "Start by understanding pheasant habitat needs," advises a Pheasants Forever pamphlet. That includes paying attention to what exists in the way of nesting areas, winter cover and the presence or absence of pesticides. "Developing and enhancing habitat . . . has proven to help increase ringneck numbers," says the pamphlet, even where hunters are present.

One late spring, out of nowhere, a ringneck pheasant appeared on our farm. There was no apparent reason for it: no one in the immediate area was raising corn or stocking pheasants. The rooster just came and stayed, his crowing alerting us to his presence even when we couldn't see his magnificent feathers glowing in the sunshine. Probably a young refugee pushed out from a farm where conditions were more appropriate, surmised a wildlife biologist who works for Pheasants Forever. It had been a good couple of years for pheasants in our general area, and some will move fair distances if they are made uncomfortable by more dominant birds in their home territories. Or, as the pamphlet says, "Because of their high productivity, 'wild' pheasants in an area can quickly populate newly-created habitat."

Wood Ducks

Wood ducks will sometimes make use of abandoned woodpecker holes, so it's no surprise that they will also nest in man-made boxes nailed to trees that grow within half a mile of sheltered water. When the hen feels that her ducklings are ready to leave the nest, she'll call to them from the ground, and out they tumble, though the nest might be as high as 65 feet in the air. The hen then marches her brood to water.

A wood duck box should be located in an open forest stand where the hen can watch for danger. A wooden box seems to be more attractive to the ducks, but a metal box keeps a brood safer from predators. A predator shield hung below the box can be helpful. If you supply the box, you'll also have to supply sawdust for nesting material because ducks don't bring their own. (68)

Gray Partridge

Gray (Hungarian) partridges that nest early can be found in open country or along the sides of roads, under fences, or along ditch banks. Later nesters often build in hay, or grain fields within 100 yards of wood edge. (126)

They eat weed seeds of plants like foxtail, wild buckwheat and knotweed, and green, leafy plants like dandelions, small grains and assorted grasses. Crops like wheat, rye, corn, oats and barley comprise a third of their summer diet. In winter, they feed on stubble and waste grain, especially corn. Young partridges devour insects like crickets and grasshoppers while insects provide only 10 percent of adult partridges' diet. (126)

Partridge habitat is an open, grassy area in a cool, dry climate. These birds prefer land that is 65 percent cultivated. Fencerows, shelterbelts and unmowed land along roadsides give partridges escape routes and nesting, loafing and winter cover. Partridges can survive in areas that are too cold for pheasants, partly because, like grouse, they can roost by burrowing into snow. (126)

Attracting Deer

Whitetail buck (USFWS)

Throughout most of the country, deer have been so successful that attracting them has become not a goal but a problem. Speaking at a symposium of ecologists, a Wisconsin botanist warned of the decline of that state's northern forests, a problem that an overabundance of deer has created. He told the group that instead of creating openings and edges, wildlife managers might have to deliberately encourage mature forests with fuller canopies that do a better job of cutting down sunlight and cutting out browse. (104)

In his warning, though, is the key to attracting deer, if, for some reason, you should want to. You should provide openings occupying 3 to 5 percent of your forest area. Each opening should be less than five acres, but large enough, or aligned so that the sun reaches most of the sod in the opening. (42)

Deer are the opposite of grouse in that they use the young, green growth in openings for food and the older forest for cover. Food within 300 feet of cover will get heavier use than food that is farther away. (42)

If tracks, trails, droppings and heavily browsed seedlings and saplings lead you to believe that deer are congregating on your property in winter, you might want to encourage them to come again by managing their wintering area or deer yard. For winter shelter, deer prefer an area of evergreens where at least half the trees are at least 35 feet tall. The trees' crowns should allow no more than 30 percent of the available sunlight to reach the ground. Maintaining such an area means avoiding clearcutting. A Vermont publication recommends that travel lanes through the area should be at least 200 feet wide. The best shelter is available in travel lanes along streams. (97)

In *Wildlife Habitat Management of Forestlands, Rangelands and Farmlands*, Neil Payne and Fred Bryant list trees important as food for deer in Maine deer yards. Arborvitae is listed as the best single browse species, while yellow birch, sugar maple, red maple and white birch are listed as having high food value when available in combination with other species. Mountain maple is listed as having high food value for deer. (89) The New Jersey Division of Parks & Forestry, Forestry Services lists sourwood, green ash, red maple, fringetree, shadbush, common persimmon, white ash, oaks and American hornbeam as trees that will attract deer. (80)

If you are determined to feed deer in winter, don't wait until starvation sets in before starting. If all other food is gone and the only thing a deer has to eat is the hay you feed it, its digestive system will not be able to make the sudden change. Once you've started feeding, continue until natural foods are again abundant. Realize, though, that artificial feeding will only result in more deer being alive to reproduce in spring — and starvation was usually the result of overpopulation in the first place. (79)

Protein-rich green plants should be available as early in spring as possible to help deer recover from the deterioration that the stress of winter causes. Forest openings on south-facing slopes where early snow-melt and green-up can occur are good places for early-greening ground cover that deer can graze. (42)

Deer's preferred browse is shrubby dogwoods, mountain and red maple, ash, white cedar, hemlock, filbert, birches, willows, mountain ash, and sumac. Acorns are often a very important food source for deer. Aspen is not a preferred food, but where it's abundant, wildlife managers work to maintain an adequate supply. (42)

Unlike grouse, deer definitely need evergreen cover in winter to reduce body heat loss. (42) Evergreens should be from 35 to 75 feet in height, with trunks at least six inches in diameter at breast height. These tall trees will provide shelter only, for they will have screened out the sunlight that promotes understory growth. This winter habitat is the most important factor in a deer's survival. Therefore, cultivating mature evergreens with large, vigorous crowns, plus a continuous food supply in nearby patch cuts of less than an acre or in food plots is the way to encourage deer.

Attracting Black Bears

A black bear's native habitat is such a large stretch of forest that almost nothing except a timber company's holdings or public lands can provide all of the bear's home range needs. However, 25 acres or less can provide feeding habitat if the land is far from civilization and in a heavily forested area. In spring, when the bear is just out of its den, it will need the new vegetation growing in wet areas. Occasionally, it will eat small mammals and carrion. Later, it will graze in secluded fields or forest openings, often eating the eggs or young of birds, rodents or other animals. Sometimes it will tear open rotten logs and ant hills to eat insects, eggs and larvae. It may also destroy bee hive boxes in an effort to get to the honey inside. In fall it will seek apples, mountain ash berries, and nuts and acorns in order to build up its fat supplies for the winter hibernation. If food is scarce in the wild, a bear may raid garbage cans. (120, 129)

In the Northeast, denning season lasts from early November until late March or early April. (99) In Wisconsin, bears are least active from late September to mid-May. (129) Den sites are most often in space dug out under a fallen tree and lined with leaves, moss or bark. Bears will also den in caves, rocks, hollow tree stumps, dense thickets, or small stands of evergreens. Once in a while in summer, a bear might shelter beside a log, tree or rock surrounded by dense shrubbery. (129)

If you seek to provide good habitat for bears, create small forest openings to encourage herbaceous growth and plant early-fruiting species of berries and cherries. Leave stump and slash to attract insects. Maintain the openings by using a brush hog or mowing so tender young grasses and forbs keep growing. (120)

LANDSCAPING FOR A VARIETY OF WILDLIFE

You can concentrate on attracting just one species to your land by shaping only the habitat that species requires, but you're increasing your chance of failure by doing things that way because unexpected diseases could wipe out the wildlife you want or the particular plant or tree that wildlife needs. A better way of approaching the goal of attracting wildlife to your land is to consider some of the general things you can do to improve a property's supply of food, water and cover, improvements that will bring a variety of wildlife to your property.

Improving Cover

Near your house, you could create a shelterbelt, an L-shaped swath of several rows of trees and shrubs that protects wildlife by giving animals and birds a place to hide in winter. The bonus for you is that it will also cut down on your heating bills. Elsewhere on your property, you could add to existing shrubs and trees to create wildlife plantings, basically large blocks of trees with many of the same characteristics as shelterbelts.

You might already have a field windbreak, a line of only one or two rows of trees whose job is to distribute snow more equally over a field. It's generally not much use to wildlife as winter cover, but birds do find places to nest in spring. (76) You could add rows of trees and intersperse additional shrubs in an existing fencerow to make a more secure travel corridor where wildlife will go from one habitat to another on your property or your neighbors'.

If your property has so much mature woodland that food for wildlife is scarce, you could create and maintain an opening in your woodlot to encourage the growth of browse that deer and other animals use, or plant a food plot in an open field near existing cover.

On a large, open area on your property — one that already exists, or one

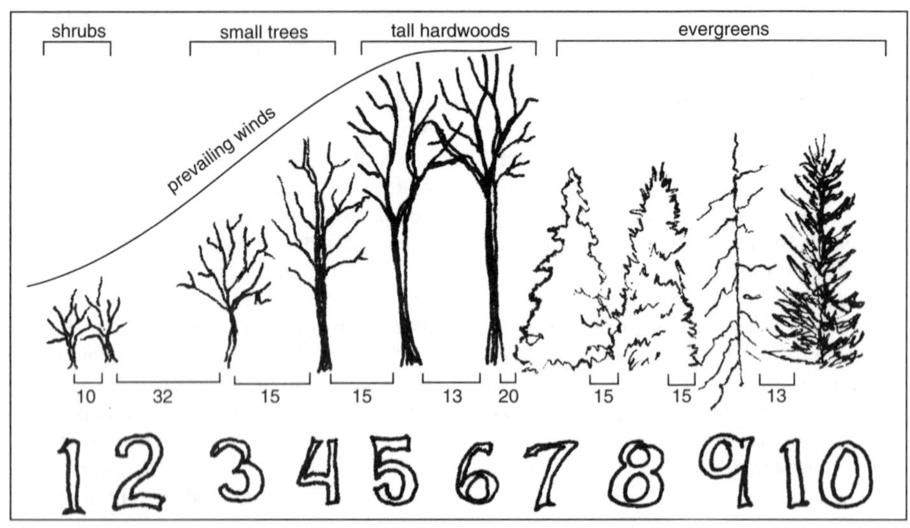

As the wind rises over a ten-row shelterbelt, snow catches between shrubs and low trees. Birds shelter beneath the wind in the interior rows of evergreens. Here, rows 7 and 8 are white pines, row 9 is Norway spruce and row 10 is white spruce. (Patricia Gilbert)

A bird seeking a shelterbelt in winter sees it this way. (Patricia Gilbert)

you create — a reconstructed prairie, complete with native flowers hiding among the native grasses and big enough for a prescribed burn now and then, would reward you with sightings of bobolinks, meadowlarks, hawks and other birds that thrive in this rare habitat.

The brief instructions included here should help you decide whether any of these are projects you want to tackle. The list of sources at the end of the book will give you more places to look for information.

SHELTERBELTS

The easiest way to create a shelterbelt is to assess what you already have, then add what is needed. A shelterbelt must have enough rows to slow drifting snow and moderate the air temperature. If it doesn't, it will be of no use to wildlife in winter. (89) The Minnesota Department of Natural Resources recommends that a shelterbelt should contain at least 10 rows of trees and shrubs and that each arm should measure at least 150 feet by 200 feet. (76) Where winters are less severe, not as many rows will be needed.

Leave any snags that you find in the area you are going to renovate as a shelterbelt, then add trees and shrubs to fill the area out and fill it in. (89)

One standard plan for a shelterbelt starts with shrubs and small trees on the outside rows to catch snow. Tall, deciduous trees are planted in the center rows to moderate the wind velocity, and evergreens are planted on the inside rows, the sides farthest away from the prevailing wind, where wildlife can shelter during a storm. Evergreens can cut the wind to speeds of below 10 miles per hour. (89, 76) A second, newer approach puts the tallest trees in the outside row and spaces the trees farther apart. In *Landscaping for Wildlife*, Carol Henderson claims that this arrangement will control drifting snow better and prolong the life of the trees in the shelterbelt. (44) The Wisconsin DNR publication, *Gimme Shelter*, calls the outer rows of shrubs optional. However, those outer rows will undeniably add wildlife cover in summer.

The inside row of the shelterbelt should be far enough away from your house to allow good air circulation in summer. (76) It should be at least 100 feet from buildings to prevent problems with drifting snow. (53) It should not cast a shadow over your driveway or a nearby road, because this could prolong icy conditions in winter. It should extend 50 feet beyond the area you want protected, but it should not extend far enough toward a highway to interfere with a driver's view or dump snow where it's not wanted. (76)

Planted in the right place, an established shelterbelt will slow the wind on the downwind side to a distance of 10 times the height of the trees. (76)

See the illustration for more information on tree planting and spacing. To choose the right kinds of trees for your area, see Table 4-1.

SHELTERBELT CARE

If you're planting a number of trees at once, hiring a tree-planting machine and crew makes sense for a number of reasons. Besides saving your back, the machine will plant much straighter rows than you could plant by hand. This will make mowing your patch of small, new trees a lot easier.

In a newly planted shelterbelt, during the first five years, watch the trees and shrubs for signs of wilting, discoloration or loss of leaves. Problems might be caused by conditions on the site like disease, insects or rodents. If you know that there are pocket gophers in your area, keep an eye out for them, because they can threaten even older stands of trees by gnawing on their roots in spring and fall. (76)

Don't prune trees that are younger than four years old unless a branch is deformed or damaged. Remove these if crowding is causing a loss of lower branches. Weeding is important at first to decrease competition, but after about five years, stop cultivating the ground around the trees. Instead, sow the area with a grass and legume mixture to give birds nesting cover and to protect the soil from erosion, and don't mow unless rodents become a problem. (76)

Some publications currently being distributed are still recommending shrubs that have been found to become invasive pests in some parts of the country. Be sure to check the status of shrubs like buckthorn, Russian or autumn olive or multiflora rose before adding more to your corner of the world.

Whichever type of shelterbelt you choose, nothing has been added, so far, to provide food for the wildlife that shelters in the trees. To remedy this, plant trees and shrubs that bear berries inside the innermost row of evergreens. A mix of berries that birds will eat as they ripen and berries that birds won't touch until winter will provide the longest-lasting food supply.

A warning: if your land is nothing but a large, open grassland, don't start a shelterbelt until you consult a local wildlife manager. Adding protection that fragments a grassland might do harm to resident grassland birds — even prairie chickens, if your grassland is in the right, rare location. (54) It's one of those upside-down decisions: forgo a shelterbelt that will decrease your heating costs for the sake of the wildlife that's already on your land.

ADDITIONAL COVER OPTIONS

Odd-shaped areas, like field corners and ravines are good places to plant clumps of 25 to 50 evergreens. Plant them 8 feet apart, so that they will keep their lower limbs for cover for as long as possible. Plant one to three rows of shrubs that produce fruit. Space these shrubs three feet to four feet apart around the edges of the evergreen plant-

In spring, wild fruit trees form a thicket that offers good cover.

ing. The shrubs will provide nesting cover and songbird food. Shrubs with thorns will also provide escape cover. Or leave half of the odd area as an opening planted with a grass and legume mixture. Plant spruce trees in 20 percent of the area and fruit-bearing shrubs and small trees in the remaining 30 percent. (89)

For instant grassy cover, simply run a fence diagonally across a corner of a pasture and let whatever is native invade. (89)

Brush is better than mature forest for providing wildlife food and cover. If your land includes a woodlot, you can improve conditions for birds and animals by providing a brushy border around its edges. The increased sunlight available at the edges of your woods probably has already encouraged new growth. Your job, then, is to maintain the stage of succession that is most useful to wildlife.

Brushy borders should be 20 to 30 feet wide. To begin, cut all trees taller than 10 feet or larger than 4 inches in diameter. Then maintain your brushy border by cutting it every 5-6 years with a brush hog. Cut only half of the border at a time so that the wildlife have some fall and winter habitat. Don't mow until late summer, after nesting season. (89)

You can also provide brushy cover by simply not removing the tops of trees and other leftovers after loggers have done their work. However, this slash can make it difficult for game animals to move around and for seedlings to regenerate if sunlight is blocked. Slash also does seedlings no favors when it gives rodents a good place to hide from predators. More surviving rodents mean more mouths feeding on seeds and seedlings. (89)

PRAIRIE AND GRASSLANDS — COVER FOR FIELD-NESTERS

For some birds and animals, "cover" is defined not as brush or trees but simply as grass tall enough to camouflage nests or this year's crop of young. Field crops and prairies both provide such cover, but after the first year, a prairie will be the easier choice, if you have no domestic animals depending on the harvest.

In order to successfully screen both a hen and her nest, any field cover must be at least 1 foot high. To check on the height of the cover you have available, have someone hold a yardstick upright. Move about 12 feet away and look. The amount obscured tells you the height of your cover. (83)

Alfalfa will provide good cover, but only if cutting can be held off until chicks are off the nest. However, this will decrease the food quality of alfalfa fed to livestock. (83)

Cool season grasses, including smooth brome grass, timothy, orchard grass, and legumes like alfalfa and clover grow best in spring and fall, when temperatures average 65 to 75 degrees. They are easier to establish than warm season grasses, but they must be replanted every three to five years. (56) If you are planting a nurse crop like flax, oats or barley that will shade weeds while the slower growing grasses take hold, mow the nurse crop to 15 inches so that it won't form seed heads that compete with the plants you want. (83)

Many ground-nesting birds, especially mallard and pintail ducks, begin laying their eggs in early April, depending on cover left over from the previous year to shelter them. Brome grass makes good cover, but heavy snows often flatten it so that it offers little or no cover in early spring. (83)

After the first season, warm-season grasses, so-called because they grow best in the heat of late summer, require less maintenance for weed control and stand upright longer in rain and snow, making seeds for food more readily available. (56) Warm-season grasses like big bluestem, Indiangrass, sideoats gramma, prairie cord grass and switchgrass are slow to start, but produce tall, dense stands of grass. Like other plants, what is native to your area is probably the best. When a prairie grass native to Missouri was tried in Wisconsin, it did not bloom early enough to produce seed before frost; when there's no seed, there's little or no wildlife value. Check on the best variety for your area with your local wildlife agency or the Soil Conservation Service (83) and see Table 5-1 for the kinds of wildlife you can attract with various grasses, grains and legumes.

When the land is being prepared for seeding, no-till planting (seeding without plowing) is the best choice to benefit wildlife and hold erosion to a minimum. When at least 90 percent of the previous year's crop residue remains on the field, food and cover are provided, and there will be less disturbance of ground nests because no-till requires fewer passes with machinery. (89)

The fall before you seed, apply a nonselective herbicide like Roundup no later than two to three weeks before the first frost to kill any actively growing undesirable plants. In spring, two or three weeks before seeding, apply herbicide again to kill any cool-season grasses and weeds that have survived. Before seeding, apply lime if a soil test shows that you need it. If nitrogen is needed, don't add it during the first year your prairie is growing because cool season grasses and weeds will benefit. (93)

Native plant seeds that are too small or have barbs or hairs that clog a conventional seeder may require a special seed drill. Some commercial seed drills are Truax, Miller, Great Plains, Tye and Marliss. Your county extension agent may be able to help you rent one of these. Some hunter organizations own such machines and lend them to landowners. Check with the ag agent about the best time for spring seeding in your area. It's important to plant seeds at the right depth: too shallow and they might not have enough moisture to germinate; too deep and they might not be able to penetrate the soil. A quarter to a half inch is a good depth. (93)

If you can't seed in spring, seed late enough in fall so that the seeds won't germinate until spring. For no-till fall

Many prairie flowers attract butterflies. (Gary Eldred)

planting, mow or spray the cool season vegetation first to reduce competition. (93)

When establishing a prairie, you should always plant at least two to six different kinds of grasses, legumes and wildflowers. This cuts down on the chance of disease wiping out your entire stand. It also increases the number of animals and birds you'll attract because of the different foods provided and the differences in the heights of nesting cover and perches for birds like meadowlarks and sparrows. Different plants also attract different insects, which in turn attract different birds. (56)

Since grassland birds tend to like a large expanse of open land, a new prairie should be planted near existing grasslands. It should be at least 50 yards from hedgerows, woodlots and dead trees to reduce the threat from predators. If you have the land and the money, plant 20 to 40 acres or more to prairie, since this is an even better defense against predators like raccoons, cowbirds, hawks, opossums and skunks, who will benefit from woody cover even more than the grassland birds would. (56)

If you're planning a prairie, do some thinking about your goals. If one of them is to recreate an area the way it was before settlers came to your neighborhood, some research is in order. An NRCS office is a good place to find historical information about native prairies in your area. Illinois residents have an excellent historical reference in *Prairie Establishment and Landscaping*.

Avoid buying seeds marked "improved" or "selected," for these will be very aggressive and will quickly overwhelm the plants you want to encourage. A stand of pure prairie grass will need 8 to 10 pounds of seed per acre. If you're trying for a prairie with lots of wildflowers, you'll need 60 percent grass seed (2 to 4 pounds per acre) plus 40 percent flower seeds by weight. If you're trying for variety, cut down on the number of seeds of downy sunflower, false sunflower, drooping yellow coneflower and New England aster, since these produce many seeds and will crowd out other prairie plants. (75) A check of vacant lots or fields in your area with a field guide to flowers in hand will also give you insight on other plants that are likely to be too successful in your personal prairie.

The first year after seeding, native plants will be spending most of their time developing roots, so you might not see much evidence of success. Therefore, what is growing should be mowed to about 6 inches to lessen competition and to let sunlight warm the ground. It's best to mow every three to four weeks to get rid of weeds and any cool season grasses that have survived. After the first year, though, mow only when ground-nesting birds are not on their nests. In Wisconsin, for instance, mowing should be done before May 15 or after July 31. (93) Mowing is necessary in the second and third years only if weeds persist. Once the prairie is established, it should only be mowed every three or four years to allow grass to accumulate and provide nesting cover. (56)

Always use a rotary mower, which will shred plants. Sickle-type mowers will smother prairie plant seedlings. If dense weeds like giant foxtail threaten your planting, mow them to 8 to 12 inches when they are 2 or 3 feet high.

Two workers control a prescribed burn with stiff rubber flappers. (Gary Eldred)

Mow sweet clover when the plants flower to keep this invasive plant in check.

An early spring burn will also help keep earlier-starting cool-season grasses and weeds from overwhelming later-starting native plants. (93)

The roots or corms of blazing star, prairie clovers, and sometimes compass plants and prairie dock can be attractive food for small mammals. If you are losing some of these plants, and have noticed that you are blessed with a healthy population of small mammals, the solution may be a fall burn to remove the winter cover the animals need.

PROTECTING BIRDS

If alfalfa or other food crops are being grown on your land, there are a number of things that you can do to protect the field-nesting birds that you've spent so much time luring to your property.

If there are streams or other riparian (water) zones on your land, field borders at least 15 feet wide should be left in these areas. (89)

Perhaps the most important practice to aid birds that nest in fields is changing the schedule for mowing those fields. Ideally, mowing should be avoided from April through mid-July (132) in order to protect wildlife that is using field grasses for cover. Birds that build their nests on the ground are not safely finished raising their broods until then.

Fawns know that their greatest protection is to lie so still that dogs or predators walk right by without noticing them. But this protective behavior also makes them potential fodder for a mower. (79)

Cadieux recommends a possible solution: mount a horizontal bar on the front of a tractor with heavy, three-foot-long chains dangling from the bar to the ground. The chains will flush fawns before the mower reaches them. (17)

If you can delay mowing your fields until mid-July or August 1, there will still be time for the cut grasses and forbs to develop enough new growth to sustain them through the winter and supply early spring cover for wildlife. If you don't get around to mowing until after September 1, leave a minimum of 12 inches of growth. (83)

Unfortunately, it is almost impossible for a farmer to wait this long to mow a productive field. Alfalfa starts losing its protein content when it ages enough to flower, and alfalfa flowers a lot earlier than mid-July in most places. Undesirable weeds from seeds form earlier than mid-July, too, and mowing is the way to stop this from happening.

A more practical possibility may be to leave unmowed patches or strips of grass in wetter areas that ground-nesting birds and deer fawns can use for cover. (132)

Improving the Food Supply

In an older woodlot, if tree branches are so thick that the canopy shades 50 to 60 percent of a space, vegetation on the forest floor will begin to disappear. What's more, heavily shaded forage will have less protein, phosphorus and calcium than some wildlife prefer. When more than 70 percent of the forest floor is shaded, no forage plants will grow. (89) If this is happening in too many places on your land, it may be time to create openings, breaks in the overstory big enough to let the sun shine in.

Small openings in a forest provide food for deer, bears and other wildlife. The annual forbs, grasses and aspen and birch seedlings attract insects that give ruffed grouse chicks, wild turkey poults and most young songbirds the protein they need. (57)

Openings are safest in hardwood forests. Evergreens do provide good cover, but raptors that prey on birds like the ruffed grouse will make use of that good cover too. (89)

Though clearcutting will provide more herbaceous cover, you might not want to strip the area bare. If there are good seed-producing trees, for instance, you might want to save them. But if you're evaluating a tree at a time of year when it is not actively producing seeds, how can you predict whether a tree is worth saving?

Research has shown that the best seed-producers among hardwood trees tend to have rounded crowns. A tree whose crown is fully exposed to the sun will produce more seeds than one that's crowded in among other trees; however, since pollen travels only 50 to 100 feet, a tree isolated from others of the same species will not produce as many seeds because it will not be as well pollinated. In general, a tree will produce the most seeds in the middle of its life. Eastern hemlock, red spruce, white pine, sugar maple, yellow birch and beech will continue to be good seed-bearers far longer than red oaks, white birch and aspen. (110)

To decide how big your openings should be, measure the diameter in inches of each tree you decide to preserve. Double the diameter measurement, then change inches to feet. A tree that is 15 inches in diameter will need an open area of 30 feet to promote the growth of shrubs that will feed wildlife.

MAINTAINING A STAGE OF SUCCESSION

If you choose a spot for your opening where the trees around the border are not mature, your newly created opening will remain in the herb-brush stage longer because those younger trees will produce fewer seeds. (28)

A spot with bad soil is a good spot for an opening, because growth will always be slow here. An area that is either too wet or too dry or where the topsoil is too shallow will be an area where growth will be hampered and suc-

The owner of this oak savanna cherishes it as a reminder of the way all of the land in her area once looked.

cession will be slower. You can locate such an area on a soil map, topographical map or forest map. Or you can look for trees that differ from the larger group. For instance, pockets of black ash, red maple, yellow birch, American elm, black spruce or hemlock growing in the middle of a stand of sugar maples signal a wet spot. (28)

To create your opening, take out all of the trees that you've chosen for removal, disc, lime and fertilize, then disc again. Remove rocks, too, to make maintenance easier in years to come. You could plant the opening with native grass seeds, or use plants like birdsfoot trefoil, white clover, rye, millet, sorghum or buckwheat. (57) Check with your county extension office to find out the best seed for your area and soil type. Hand-broadcast the seed in early spring to take advantage of frost heaves. (43)

To keep the area open, run a brush hog through the opening every other year in mid-summer. Earlier, and you will kill birds that are nesting in the grass. Later, and the grasses and forbs will not have time to grow back before fall. Once grasses have begun to dominate, burn the patch every two or three years to slow succession.

You could also encourage the invasion of native herbaceous plants on sites where there is no serious erosion prob-

Oak seedlings sprout naturally as soon as the forest canopy is opened.

lem or a lot of woody plants like aspen that sprout easily. Just disturbing the opening with a tiller, or with the bulldozer that's removing trees, will expose the soil to native seeds. (43)

Removing trees with thick crowns will encourage browse to grow, but it will also allow smaller, more valuable trees to flourish in the increased sunshine. For instance, trees that produce edible nuts and fruits (mast) — and oak heads the list — are more valuable to someone trying to encourage wildlife than some of the taller, timber-producing trees. Ideally, a woodlot of mixed trees should contain about 40 percent mast-producing trees. (89)

If you're creating an opening by cutting trees, you might as well create snags at the edges while you're at it. Simply choose a few of the trees with a diameter of six inches or a little more. Instead of cutting them down, take an ax or chainsaw and cut two circles or a continuous band through the bark and into the cambium layer beneath it. Cut these "bracelets" a foot and a half or two feet from the ground. Within two or three months, the still-standing tree will be dead, its shade-producing leaves gone, and well on its way to providing homes for woodpeckers and other cavity nesting birds.

The longest-lasting snags will come from yellow birch, sugar maple, elm, oak, and white pine. These might stand for decades. First, the tree's small branches and twigs will fall; later, the big branches and tops will decay. Basswood and ash snags will remain standing a shorter time, while beech, white birch, and red maple snags will remain standing the shortest time. (110)

FENCEROWS

If there was a fence on your property at one time, or if there still is, birds have probably provided the start of a fencerow, another kind of habitat that you can bolster to help wildlife. (37) Here's what happened: the birds found berries and seeds elsewhere, ate them elsewhere, then sat on your fence and relieved themselves of the parts they couldn't digest. Some of those parts contained seeds that sprouted, and some of those sprouts grew into a tangled thicket that doesn't look tidy, but is the start of an animal shelter nonetheless, not only one that provides places for animals to rest, but also one that provides a safe highway for animals that are travelling from habitat to habitat.

If cottontail rabbits are your beneficiaries of choice, the single row that probably already exists will not provide enough cover. Your fencerows should be at least three rows and 30 to 50 feet wide. (43) If you add small shrubs, you'll be providing a safe travel lane. (57)

One year, the rabbits on our property taught us what works for them. Where a fence separates a cleanly cultivated farmer's field from a township road whose grass borders are regularly mowed, the only cover grows in a single row right along the fence. But the road dips at the property line, the fence makes a right angle, and several rows of brush have sprouted in the patch on our side of the fence that has been unused for at least eight years. That spring, we rarely saw rabbits along the fence, except in the thick shelter in the dip at the property line, where they routinely spurted out whenever a car seemed to be threatening them. That year we also learned the effect of live-in predators: for a number of years, a red-tailed hawk had patrolled the fencerow, but the year the rabbits were very evident, the redtail was nowhere to be seen.

Once widened, fencerows should be cut to maintain their food and cover value so that they don't grow too tall for small mammals like chipmunks and gophers to reach and browse and too leggy to make good hiding places for pheasants seeking cover when surrounding cropfields are mowed. Cutting will stimulate new sprouts, release mast-producing trees and shrubs, and increase vertical diversity. As shrubs like ninebark, highbush cranberry, nannyberry and dogwood begin to grow and wild grape, butternut and Virginia creeper move in, songbirds like cardinals, chipping sparrows, brown thrashers and catbirds will begin to build their nests in your fencerow. As with the brush bordering woodlots, fencerows should be cut in rotation, so that there will always be some cover for the animals that have come to rely on finding it. A slightly different option is to deliberately leave good food-producing shrubs but cut the rest back. (132, 54)

In addition to lines of brush that provide travel lanes, you can provide brushpiles for cover. These should be located near food plots, unused areas, other rock and brush piles, and areas that function as wildlife corridors: fencerows, hedgerows, field borders or shelterbelts. (89)

Brushpiles can be made from slash, unsaleable logs, old telephone poles, old fence posts, or trees. (89) If you have some slash that is piled high and some that is scattered, you will, of course, be creating varied habitats, with the accompanying possibility for diverse wildlife. Be aware, though, that slash rots faster if it is in contact with the ground or is left on wetter, warmer sites. The slash from eastern hemlock and other evergreens might last for decades, but the slash from hardwood trees will last only five to eight years. (110)

For more on creating permanent brushpiles, see Chapter Four.

SPECIAL TREES

Apples and acorns equal "food" to more species than any other food sources. Obviously, encouraging the trees that produce them is a good way to bring wildlife to your land.

Apples or the seeds from wild apple trees provide food for deer, ruffed grouse, hares, rabbits, squirrels, foxes and many other species of wildlife, though apples are also favored by animals you might be less happy to see, like porcupines and bobcats. Apple trees also provide summer habitat for songbirds.

Wild apple seeds are carried in bird droppings. They tend to grow in clearings or at the edges of fields. (36) Overgrown apple trees also can be the result of abandoned apple orchards. By the time you discover these trees, succession will probably have added shrubs around them, and perhaps other trees will have grown even taller than the apple trees, overshadowing them.

If you have some big, old apple trees on your property, it's obviously faster to recondition them than to plant new ones

(Patricia Gilbert)

and wait until they reach this size. You'll have to work on two things, the apple tree itself and its neighboring competition. On the tree you want to save, remove dead and diseased branches, and all but the largest, healthiest stem. Remove about one third of the remaining live growth, opening up thick clusters of branches and clipping one to two feet from the ends of side branches and vertical sucker shoots. Also remove any shrubs growing within the drip line of the apple tree and any trees that are shading out the apple tree on at least three sides of it, especially the south side. If a soil test shows that fertilizer is needed, apply it in a narrow band right under the drip line so that the feeder roots will benefit from it as it seeps into the soil. (36)

If a large tree must be removed to "release" the apple tree, simply girdle it and create a useable snag. Smaller trees can be removed and their stumps sprayed to keep them from growing back, but don't spray the stumps of surrounding apple trees, for their roots might be connected to the tree you're working so hard to preserve. (36)

The one exception to completely removing undergrowth is for grouse. If dense evergreen or brush cover is growing close to an old apple tree in grouse territory, leave this escape and roosting cover on one side of the tree, the north side, if possible. (36)

Oaks produce acorns for many species and roosts for turkeys. Large, broad-limbed, open-branched oak trees on south- or east-facing slopes, especially those higher up on the slope, are good candidates for turkey roosts. Such trees are worth preserving if you're in turkey country, and more and more people are.

Oak seedlings may be available from your state tree nursery. Certainly they will be available from commercial nurseries. But you can also start oak seedlings the way Nature does, by gathering acorns as the nuts fall off the trees. Put these fresh acorns in water and discard

A plastic mesh sleeve protects a newly planted oak seedling against hungry deer.

any that float. Then plant a lot in each place you want them to grow to give squirrels and other wildlife some to eat, because they will anyway. The same rule applies here as it does for maintaining openings: the better the soil where you plant, the harder it will be to keep out competing trees like maple, basswood and elm and competing shrubs like blackberry.

Growing your own seedlings is cheaper than buying them, but it is also going to be slower. One professional forester experimenting with planting acorns found that at the end of the first growing season, the seedlings were less than a foot high. Close by, during the same growing season, she had planted two-year-old seedlings. These were about five feet tall by the time fall arrived.

Since deer are partial to oak seedlings, ones that you grow or ones that you buy should be protected if you live in deer territory. A small area of new trees can be surrounded by a chicken wire fence, though scattered seedlings will need individual protection. White oak seeds can be replanted as soon as they fall, though red oaks can be planted the next spring if they're kept in cold storage during the winter. As with any other young trees, weed your acorn sprouts to protect them from competition. (70)

Don't prune young oak trees in late spring through early summer (May through July in Wisconsin) when they are most likely to be attacked by oak wilt, a fungus spread from infected trees by insects, spores, or connecting roots. Red oaks are more susceptible to oak wilt, but the disease attacks white oaks too. (70)

It goes without saying that when your woodlot gets old enough to cut, oaks should be left if you're interested in wildlife. At least a quarter of the oaks should be at least 75 years old. The ones on the upper slopes will not only be favored by roosting turkeys, they'll also be less prone to frost damage. (70)

FOOD PLOTS

If wild turkeys, pheasants or grouse are on your list of things you'd like to see more of, or if being kind to animals is on your to-do list, a food plot or two might make a good addition to your property, as long as it is close to a sheltered area, and on the shelter's leeward side. A food plot located near the inside of a shelterbelt will keep animals in good physical condition during a mild winter and might mean the difference between life and death in a hard winter. (40) Quail and wild turkey hens who haven't been able to find enough winter food won't lay eggs as soon. If their first nest is destroyed, they won't have time to lay a second clutch of eggs. Besides, lightweight hens are much more likely to die during molting season from heat or disease, and their chicks will also have a higher mortality rate. (54)

Unlike a feeder, which has to be built or bought, filled, and replenished, a food plot is a small field planted in spring that is left unharvested in fall and yields seedheads that will feed wildlife through the winter. A food plot can also contain food that is meant to be eaten as soon as it is ripe.

As noted above, corn is the most valuable grain to plant, because it stands so tall and continues to stand tall above the winter snows. Millets and sorghums that will remain standing in winter are also good choices. Buckwheat has weak stems, but deer will probably finish it off long before winter comes. (40)

Perennial crops like clover, alfalfa and other legumes make summer meals for turkeys, songbirds, rabbits and deer. Small grains like millet, oats, wheat, rye and buckwheat will provide food for songbirds. (54) A food plot of sunflower seeds will provide food for birds and small mammals. (132)

Your food plot should be wide enough so that it does not simply act like a snow fence, allowing the wind to dump snow until it covers the food.

One blizzard might fill as much as the outside 25 to 50 rows of a patch of standing grain sorghum or corn. A Pheasants Forever publication says that for this reason, a large food plot, as large as three to ten acres, is the most desireable, and a square patch is better than a long, narrow one. If you don't have room for a patch this big, establish a snow trap by harvesting 12 to 20 rows just inside the outer four to six rows on the windward side of the patch. (70, 91) You'll be sacrificing four or six rows, but the remaining rows should stay relatively clear of snow.

Payne and Bryant advise that a 10 to 30 foot wide strip of unharvested crops makes a good food plot. The rows should be close to the shelter of brush or trees so that animals and birds that feed there do not have to cross long, unprotected spaces in order to make use of the food. The plot should be planted in a long, narrow rectangle to increase edge. If your food plot does not seem to be doing well, you may have to root-prune the neighboring woody growth every four years. (89)

One wild turkey will eat half an ear of corn a day. This means that 12 rows of corn in a food plot that is 50 feet long will support 20 turkeys for three months. (132)

Because a food plot usually covers at least an acre or two, hand-broadcasting the seed is not practical. If you don't own a tractor and/or the right type of seeder, your state department of natural resources or your county extension agent may be able to help you to find someone who can plant the seed for you. You can also advertise in a local newspaper, or just ask around the neighborhood to find out who does what is known to farmers as custom work.

Various organizations collect outdated seed from agricultural suppliers and make it available free of charge to anyone who wants to plant a food plot. For instance, Wings Over Wisconsin makes this an annual spring project, and for some local chapters of Pheasants Forever, distributing outdated seed is the project of choice.

If you're successful in drawing wildlife to your food plot, you might have to re-plant the plot every year. (117) However, if you leave a food plot idle for three years after it is planted, early succession plants will invade, attract insects, and provide additional food for songbirds and other wildlife. (132)

Another plan for a food plot suggests alternating strips of crops with strips of natural vegetation that have been mowed and left idle so that native grasses and weeds invade. (118) Lightly discing an unused field will also encourage annual seed-producing weeds. The exception is a remnant prairie, a type of unused field that is too rare to plow under. (70)

A food plot should not be planted too near areas where there is a lot of traffic, both for the sake of the animals who might bolt onto the road and the drivers who might hit them. It should be planted in an area that is not likely to erode. If it is planted on a slope that is steeper than a five percent grade, it should be planted in contours.

If your land is enrolled in the federal Conservation Reserve Program, you will not be allowed to plant soybeans or sunflowers. (116) You will, however, have to clear the location with the Natural Resources Conservation Service (NRCS), and plant no more than 10 percent of CRP acreage in food plots. You'll also have to protect your food plots from grazing animals and reseed a vacated plot with approved cover. (117)

Use herbicide only if noxious weeds threaten to become a problem. Remember, the object of this plot is not to harvest as much grain as you can sell; the object is to feed wildlife. And wildlife does make use of weeds and the insects they attract. (116)

A food plot could increase your opportunity to observe wild animals and birds where you don't otherwise see them. However, like bird feeders, food plots do tend to concentrate healthy and sick animals in an area where predators can find them. (89)

CALCULATING SUCCESS

How soon can you expect to see an increase in the wildlife population on your land? Loggers in northern forests report that deer will appear in response to the snarl of their chain saws to munch the tops of just-felled trees. (42)

But a lot depends on what a bird or animal is looking for in the habitat you provide. Sparrows and wrens may come almost at once to make use of logging leftovers, but other songbirds may wait until new sprouts have begun to branch and provide nesting sites. Cavity nesters will definitely wait until decay sets in, making a tree easier to excavate. And raptors arrive in the second act, after the animals or birds they prey on have made themselves at home. (42)

But a prey species will survive even hunting pressure, as long as the habitat is maintained. "Ducks are hunted, yet species survive, in part because of the support they get from duck hunters interested in maintaining their sport," says *The Birder's Handbook*, a publication that is generally neutral on the question of hunting.

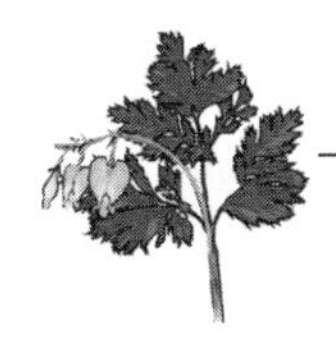

Chapter 6

FINDING BALANCE

You can't be too rich or too thin, someone once said. Someone named Vanderbilt, I believe. And then there's the country song that says you can't have too much fun.

But can you have too much wildlife?

There was a time when I would have said no way.

I recall standing as a young boy on the Wisconsin shore of Lake Michigan and staring in awe as the very sky seemed in motion, filled with untold thousands of ducks fleeing south on the north winds of a cold front. Too many ducks? No, just about right.

And then, later in life, standing on the shore of a river in the far north of Quebec. A dozen of us caribou hunters had spent our blood lust in the few previous days, and meat poles sagged with quarters of venison. Now we simply watched in hushed silence as a seemingly endless line of caribou bulls approached across the far tundra and then began swimming the river. Only their heads and their incredible antlers were visible above the water, and it was like watching a moving line of mesquite brush. As they reached our side of the river, each caribou stepped from the water, shook vigorously like a dog, and marched onward. Too many caribou? No, just about right.

And then other times, standing with Marilyn on the shoulder of a highway that crosses the Horicon Marsh in east central Wisconsin, waiting for the setting sun to touch the horizon. At that signal, Canada geese rose from corn and alfalfa fields from miles around and headed for the watery refuge of the great marsh. The excited calling of nearby geese raised an ear-splitting din, and the flocks approaching were like clouds of smoke across the entire northern and western horizons. Too many geese? No, just about right.

At the time, we owned just a half-acre of suburbia in Milwaukee County, and we delighted in seeing songbirds and deer tracks in our own back yard. The only problem with wildlife was more of a joke. Raccoons would rattle the garbage can lids deep in the night and our son, Tony, would wake up, ease his bedroom window up a few inches and carefully aim his BB gun. A ping and a yipe later and the yard would be silent again and Tony would fall asleep with a smile on his face. We'd all have a good laugh at breakfast.

It didn't really dawn on me that there could be too much wildlife until I went to hunt geese on a farm not far from that Horicon Marsh. The farmer, a businesslike man named Tom, welcomed me as a stranger, showed me where to put up a hay-bale blind, and seemed genuinely pleased when I shot a goose the next morning.

And then as we talked, I discovered why Tom was such a nice guy. The numbers of geese had exploded in recent years, and they were making a shambles of some of the prosperous farm that Tom had built up with so much hard work. Not only were the thousands of geese reaching up to steal kernels from ears of corn, but they were working over his alfalfa fields. Tom showed me how a goose would grab an alfalfa plant and give it a yank. That gave the goose a beakful of nutritious legume leaves, but in the process it also broke the bond between the alfalfa roots and the soil. The plant then slowly died. Acre after acre of alfalfa was browning and dying. The following spring, Tom would have to replant this crop that should normally last for five to seven years.

The federal government had supplied Tom with propane-fueled cannons that went off every few minutes with a report like a shotgun. But geese quickly learned that the cannons were shooting blanks. Unless, that is, they saw a goose fall fluttering to the ground now and then. That made believers out of the remaining geese, and that's why Tom was so nice to me and several other hunters. We kept those geese on the move and helped Tom salvage part of his crops.

There are those who don't like the idea of killing animals for sport, and I have many friends among that persuasion. I don't argue with them unless they bring it up, but then I don't hesitate to state my personal views. Those include the deeply-held opinion that I am a child of nature myself, just like the fox and hawk. If it's all right for the fox and hawk to kill in order to eat, then I claim the same legitimacy for myself. Like them, I don't kill anything just to kill it. A wild turkey is the centerpiece of our Thanksgiving table, trout and bluegills are eaten to the tune of stories of how they were caught, venison is a special meal at our house, and when I jump on my Harley I'm wearing a black jacket made from the hides of deer that I have killed.

Since retiring from 30 years of newspaper work, I've become a farmer, raising hay and pasturing beef cattle on 160 acres of hilly land in southwest Wisconsin. I've also become active in conservation, hunter and farmer organizations.

Farmers have shown me cornfields in which they must write off the outside four rows, losing every ear to marauding deer and raccoons. I've been on committees that wrestled with the question of how to compensate and protect farmers, and who should pay the very steep bill.

I enjoy the howl of a wolf on a frosty night as much as anyone, but it's unrealistic to think that wolves will ever be back in the dairy- and beef-farming areas of the Midwest in numbers that would control deer. But deer don't know that, and they reproduce at the same rate they did when wolves took every second fawn.

If human hunters, including me, didn't take a carefully-regulated share of deer, the numbers would explode to the point that farming would not be possible. This is not just a personal opinion; it's based on my reading of much research on the subject.

On another level, word spread one summer among local raccoons that delicious chicken dinners were free for the taking in the Leys barn. Well, I'm the protector of those chickens, a job that I take seriously. I tried live-trapping coons and setting them free somewhere else, but that didn't work very well. So I used a combination of leg-hold trap and .22 rifle. Fifteen raccoons later, the message seemed to have gotten around. That was several years ago, and I have had very little trouble since.

It's never easy to shoot an animal in a trap, so I always brace myself by saying that the raccoon was about to eat a chicken in a pen.

One recent summer, I thought it would be nice to have a couple of turkeys and a couple of geese to wander the yard, penning them up at night like we do the chickens. You have no idea, until you do that, how much poop comes out of a goose in a day's time. No, more than that. Geese eat green grass, just like cows, and most of that grass comes right out the other end of the goose in a form that sticks to Marilyn's shoes and just about everything else.

Golfers around the country have discovered that, and so have owners of those lovely condos that surround those lovely suburban ponds. They have also discovered that geese can be noisy, disagreeable creatures, taken to loudly chasing children and dogs and cats and anything else that annoys them.

Deer have multiplied beyond their welcome in many suburban neighborhoods, reproducing in the absence of even human predators. City and village boards wrestle with the question of reconciling desires of auto insurers and gardeners with those who can't seem to ever see enough of those lovely deer.

The tide often slowly but surely shifts. Live-trapping is tried, but usually found to be an expensive process that accidentally kills many deer. Other means are discussed, sometimes loudly and angrily and late into the night at public meetings. I recall sitting in a bar near Milwaukee one night and opining to a state wildlife manager that suburbanites would never have the stomach to just kill some deer. I ran into that wildlife manager some years later, long after I had left the Milwaukee area. He said, guess what, they're killing deer in Milwaukee County. Hiring sharpshooters to kill them at night over floodlit piles of corn.

In and around Des Moines, Iowa, government officials have responded a little differently to complaints that deer were ruining the parks. They now allow bowhunters to take some deer, under carefully controlled conditions. The difficulty with bowhunting is that even a mortally-wounded deer will often travel some distance before dying in the back yard of a surprised little girl.

On a more personal note, a couple of friends decided that, by God, they were going to finally have a vegetable garden that wouldn't be eaten up by raccoons, chipmunks, rabbits, woodchucks and other lovely creatures. They spent untold hours and dollars erecting a permanent rabbit-wire and electric-wire fence around the whole garden.

After they were done, I opined that it would have been cheaper and a lot less trouble to just get a good dog. It seems they hadn't thought of that.

— Ron Leys

TOO MUCH OF A GOOD THING

A piece of land that provides the right kinds of food, water, cover and space will attract the type of wildlife that needs that combination — assuming, of course, that the wildlife is in the area to begin with. But sometimes a landowner's luck is too good and too many of a species settle on a property.

What is meant by "too many"? If wild animals are eating themselves out of the very food they need, or exhausting their water supply, they have exceeded the biological carrying capacity of your land for their species.

Social carrying capacity or landowner tolerance is another way of looking at the question of "too many." This has to do with the fact that we humans will often not tolerate as many of a species as the habitat can provide for; in some cases, even one skunk is too many.

If the habitat you have created becomes too attractive, your first task is to identify exactly who the problem birds or animals are. This is sometimes not as simple as it sounds.

Take the case of wild turkeys. Besides eating waste grain and crops that haven't been harvested, turkeys will also take advantage of crops that were first damaged by deer, squirrels, raccoons or crows. When wildlife managers in Wisconsin investigated complaints of crop damage caused by turkeys, they found that most of the damage was actually caused by other wild animals that arrived first, but simply were less visible or did the damage at night. In one Wisconsin study, of 28 complaints of turkey damage investigated, deer caused 55 percent of the damage and raccoons caused 25 percent. (70)

Once you've identified the culprits, you can try to deal with them in two different ways — you can remove them, or you can do your best to keep them off your property, or away from those parts of your property where they are causing the problems.

Intruders can be removed, if legally allowed, by live-trapping and transportation to a remote area near appropriate food and water, where their presence won't bother either the animals who already live there or any human inhabitants. There are problems with removing animals, however: research has shown that some removed animals have remarkable persistence and will travel amazing distances in order to return to what they consider home. Also, opinions vary on what percentage of removed animals will die because of the stress caused by the removal. Deliberate killing — shooting or trapping, or poisoning where legal and where it won't put other animals in danger — is a far more permanent solution, though also a far more controversial one.

You can also try to get rid of the intruders by inverting the advice that you've read in this book: figure out which features of your habitat are drawing the intruders; remove those features, and let the animals take care of removing themselves. Or you can try using repellents or fences. (134)

Is it selfish to try to keep wildlife away from your back yard? That's a personal decision. But if you have decided that enough is enough, there are ways to deal with your problems.

Geese

Some years ago, a new race of Canada geese began turning up in urban areas around the country. They were bigger than the migrant geese that residents were used to seeing only in spring and fall, and they hung around all year. They quickly learned how to enjoy life in areas where there was a shortage of natural predators plus many patches of short grass on lawns, in parks and on golf courses. As for the humans the geese encountered, instead of being hunters, they were folks who often came bearing food. Ironically, the areas where human access to the geese was easiest — sidewalks, lawns, parking lots — soon became the messiest because of the concentrations of freeloading Giant Canada geese on them.

Canada geese. (USFWS/Wyman Meinzer)

The new geese have prospered. Because they aren't exposed to hunting, these urban geese have a higher survival rate than geese that migrate to and from northern Canada and stop off at rural wetlands on their way. Fifteen- or 20-year-old urban geese are common. They also tend to breed at a younger age. (73) Moreover, they might nest within 6 to 10 feet of each other, while most migratory geese won't tolerate such crowded conditions. (105) If your urban property is near water, you are probably well aware of the resulting population explosion.

Knowing the life cycle of geese and habitat needs will help you understand what might work if you want to repel them. Watching to see whether they habitually fly or walk into an area will also help you select methods to try. Unfortunately, many of these methods fall into the category of You-Can't-Win.

Geese are partial to young, tender grass shoots that are high in the proteins and carbohydrates they need. (73) If you maintain a lawn with grass that is 6 inches tall, and stop watering during dry periods, you'll reduce the supply of young, tender shoots. There's some evidence, too, that geese prefer fertilized to unfertilized grass, so cutting back on fertilizer is another possible partial solution. (105)

Given a choice, geese like Kentucky bluegrass more than tall fescue, but they'll feed on almost any short grass or legume, including brome grasses, new growth on canary grass, colonial bentgrass, perennial ryegrass, quackgrass, red fescue or new growth on switchgrass that has been mowed or burned. You might want to re-plant strategic areas with plants that geese don't care to walk through or eat. Besides tall fescue, they'll avoid periwinkle, myrtle, pachysandra, English ivy, hosta or plaintain lily, euonymous fortuni and ground juniper. (105)

The Environmental Protection Agency has approved ReJeXiT and Goose Chase for use on lawns, though not in water. These chemicals make grass taste bad to geese, but that won't keep them from walking over it to get to other, untreated areas. (73)

There are, of course, a number of places where longer grass is not an option. Besides, geese may not like to walk through tall grass, but they might use it for nesting sites. (105)

Geese feel most protected on water. When they feed on land, they want an uninterrupted view of the area between themselves and the water. One idea for keeping them from walking onto your property from the water that borders it is to modify an uninterrupted shoreline by putting shrubs or boulders every 10 to 20 yards in order to spoil the geese's clear view. Boulders must be at least two feet in diameter to be effective, and shrubs should be dense and at least 30 inches high so the geese can't see through or over them, or walk through them. The downside is that geese will sometimes find such shrubs and hedges to be ideal nesting sites. (105)

Cattails, bulrushes and other tall aquatic plants planted between water and grass might prevent geese from coming ashore. However, these plants may favor muskrats, which can cause problems of their own. (105)

A walking path between water and grass might interrupt the geese's access. Unless, of course, humans are using the path as an easy way to get to the geese and feed them. Geese are opportunistic feeders, meaning that they'll eat whatever is easiest, including bread and popcorn, which doesn't provide very good nutrition for them. (105, 73)

Usually, urban geese will nest within 150 feet of water. In Virginia, they will nest in late February, in northern Illinois, in late March to mid-April and in central Wisconsin, around April 4. They'll choose open areas with a clear view of predators, and they're especially fond of urban lawns next to water. (105)

It might seem that the easiest way to control a population of geese would be to remove eggs before they hatch. Unfortunately, this will accomplish nothing, since birds will just lay more eggs until they have what they consider an adequate number. However, if eggs are left in the nest but interfered with, more eggs will not arrive. The Cornell University Extension's technical guide suggests oiling, shaking or puncturing eggs, or replacing the eggs after the first week of incubation with wooden or hard-boiled eggs. It suggests that a team try this, with one person assigned to chase off the defending gander and one to handle the eggs. It also warns that before carrying out any of these procedures, you should check with federal and local officials to see what is allowed and whether permits are necessary.

If geese are landing where they aren't wanted, it might work to haze them while providing a field with even better forage in a rural or suburban area where people will tolerate a few geese. To lure them, plant grain crops, then cut or flood the area when the grain is ripe. Or provide a well-fertilized, mowed site of Kentucky bluegrass. (105)

However, if you manage to attract lots of geese to your alternate site, trampling will make it less attractive to the next wave of geese that comes along. Clover resists trampling better than row grain crops, but nothing is perfect. Another option is to provide loose grain at a bait station to minimize trampling and feed more geese. (105)

Molting — when old feathers fall out and new ones take their place — takes about a month. The period begins on June 1 in northern states and June 15 in mid-Atlantic states. Non-breeding yearlings, non-nesting adults and adults whose nests were destroyed will be the earliest to molt. If the presence of geese is going to annoy you, this may be the time of year it happens, for this is the period when geese can't fly, so they produce manure in the most concentrated quantities. (105) Fences or string barriers that geese won't walk through work best during the molt, but they should be installed early in the year when geese are just starting to invade. Low wires work best, but some people have had success using two 20-pound-test fishing lines placed 6 and 12 inches above the ground and staked at 6-foot intervals. These must be checked often, though, and entangled geese freed. (73)

Before you try to control geese by doing something other than changing your landscaping scheme, check with your local USDA Animal and Plant Health Inspection Services office about the legality of what you plan. Also check with local authorities and consider the impact on your human neighbors. (105)

If geese are flying onto your property, or your property is larger than several acres, the only thing that works is harassing the geese with such things as plastic construction flags or maypole streamers, carbide or propane cannons or harassment by dogs or humans. But according to the Maryland Department of Natural Resources publication on the subject of goose control, "Hunting is considered to be the most important management tool for controlling local Canada geese populations." The publication is not advocating the death of every goose on a property. It is merely suggesting that on a property where noisemakers are being used, when some geese die, others will get the idea that the noises they hear could kill them.

Geese are very bright animals that quickly become used to whatever method you are using to scare them. Using more than one method at a time and moving the repellent devices often may help.

If you couple scary sounds with a human scarecrow that has moving arms, you increase the chance that the scarecrow will scare. This method is best if the scarecrow is in place before the geese arrive. (105) Geese also don't like to hang around if there's an object hovering overhead. Some people have had some success with mylar tape that reflects sunlight and produces a humming noise when the wind stretches it. Flagging in bright colors or orange construction flagging can be hung from trees. Thirty-inch helium-filled balloons — especially those with eyespots — can be tethered on 10 to 30 foot long lines of 50- to 70-pound-test fishing lines. However, you may need to move these objects often as the geese get used to them. (73)

By federal law, dogs may be used to harass geese, but not to catch them. Border collies and other herding dogs work best. At first, geese must be chased several times a day for several weeks. Then you'll need to patrol the area regularly, but less frequently. This is one method to which geese do not become acclimated. But it shouldn't be used when geese are nesting, when goslings are present, or during the molt when the birds are unable to fly. Dogs will keep geese off lawns, but not off the water, where the birds have the advantage. (73)

Noisemakers are also options for harassment, but be sure to check with local police about noise control ordinances, fire safety codes and restrictions on the possession and discharge of guns before using such items as sirens, bird bangers and whistles, bombs fired from hand-held 15-millimeter pistols or shell crackers — firecracker shells fired from a 12-gauge shotgun. (73)

Even if noisemakers are legal, there are other considerations. One year, one of our neighbors decided to protect his wine grapes from songbirds by renting a noisemaking cannon. It did not make much of an impression on the birds, but it made quite an impression on us — and we live more than half a mile away.

Actually, the best defense against having a goose problem is not to give in to the impulse of feeding the first pair that arrives on your property before or during breeding season. Realize that even with a single pair, the gander might become aggressive in defending his young. As the young grow, they will add their waste to the adults'. Then one couple might attract other couples, and the problem accelerates. To keep the first pair away, build an 18-inch high fence of string between your lawn and the nearby lake or pond that is offering the geese shelter. According to the author of *Nature from Your Back Door*, geese won't cross the string, especially if the goslings are too young to fly, or during the molting season. (34)

If you want additional, very detailed information about techniques for trying to get rid of an overabundance of geese, check out *Managing Canada Geese in Urban Environments. When Resident Geese Become a Problem* on the Maryland Department of Natural Resources Website, lists methods that have not worked and sources of additional information and supplies for repelling geese.

Other Problem Birds

Sometimes birds will build their nests right where you wish they wouldn't — over or near the door you use most often to get in and out of your house, for instance. One problem is that bird droppings land where no one can avoid stepping in them and sometimes land on someone who had no ambition to be a target; a second problem comes when an overprotective parent starts dive-bombing your guests. A long pole with a hook on it is useful for pulling down unwanted nests, but you'll have to repeat the process about every two weeks. (134)

Where small flocks of roosting blackbirds or starlings are the problem, noise may help. It should begin as birds begin to flock onto trees or onto buildings and continue until the last bird leaves. Start early in the evening, while there's enough light for birds to find an alternate roosting place. Hollering, clapping or beating on pots are all worth trying. For larger groups, gas exploders or the recorded calls of starlings in distress might work. Extremely large flocks require professional advice. The U.S. Fish and Wildlife Service might be able to help. (134)

If birds are helping themselves to more of your fruit and berries than you care to lose, plastic twirlers, cloth strips, aluminum foil strips or other moving objects might work in a small area. For a larger area, noise techniques might work. Of course, if it's just one tree at a time that is being robbed, netting will be the answer, as long as the birds can't reach through it or fly underneath it. (134)

Insect-eating woodpeckers can be a help, but sometimes they get overzealous. Regular rows of woodpecker holes in the siding of your house show that birds were looking for insects. Holes about 1 inch in length may be round and smooth or vertical and rough. The problem comes from the fact that moisture, decay and insects can enter these holes. The trick is to begin trying to scare it away as soon as a woodpecker becomes a pest. (19)

It might be possible to scare the bird away by tacking aluminum foil or plastic streamers to the damaged area. These will move in a breeze and might frighten the bird. (19)

If an overabundance of insects is attracting the woodpeckers, wood preservatives might repel the insects and the woodpeckers as well. Sticky, paste-type repellents that would stain wood should be applied on masking tape put over the wood. (19)

What songbirds accomplish by singing in spring and fall, woodpeckers accomplish by drumming. But if incessant drumming is driving you crazy, find the site that the woodpecker is using and cover it with padding, or provide an alternate site with two boards secured together but still resonant. Or fill in all hollow spaces to deaden the sound. (19)

Sapsuckers cause a special kind of damage. Especially during the spring migration, these woodpeckers might return repeatedly to the trunks and limbs of ornamental or orchard trees. The damage here is to the tree, which will show less growth and increased disease and insect damage. Protect such trees by draping plastic or burlap over areas damaged by many rows of holes, or wrap loose cylinders of the wire mesh known as hardware cloth around the trees' trunks and limbs. (19)

Deer

By April 25, "the ground in the woodlands should be blanketed with spring ephemerals such as toothwort and trout lily," the printed guide to the University of Wisconsin-Madison Arboretum promises. In the park itself, some trout lily leaves are showing their liverish spots, but there are no flowers. In fact, almost nowhere in the late-April woods are there blankets of flowers, except in a few of the more open spots. The problem? Deer.

Deer are on the mind of a volunteer pruning roses. Growing roses here is a lost cause, he says, unless they are the old-fashioned rugosa roses with thorns so vicious that even a deer won't touch the flowers. The other roses get browsed all summer. Each yew, hemlock and arborvitae (white cedar) has its own mesh fence secured by fenceposts. Young trees have wire panels attached to concrete-reinforcing rods. The panels start about a foot above the ground and surround the trunks. This is to keep bucks from rubbing their antlers on the trunks of the supple young trees. When the trunks are about 4 inches in diameter, the deer will lose interest and the fences will be removed.

In late April, in the southwest corner of Milwaukee County, at the Boerner Botanical Gardens, wherever a bed of tulips hasn't bloomed yet, there's a three-

It's spring, but the snowfence is up in Milwaukee County's Boerner Botanical Gardens. This fence is separating tulip bulbs that have yet to bloom from the park's deer.

Flapping grocery bags keep deer from feeding on unopened tulip bulbs -- at least for a while.

foot-high stick in the center of the bed, and on every stick a white plastic grocery bag is tied by its handles. When the wind blows, the bags fill, and stir restlessly on their sticks. The bags are no less ugly than the circles of snow fencing that border some of the beds, or the green plastic mesh that covers others.

The receptionist at the park's office is enthusiastic about the flapping bags. The park's deer hate the sound, she says, and if the sound doesn't scare them, the motion does. Julian Westley, the park's head gardener, is less optimistic. If the deer were just passing through, the bags might scare them, he says, but these deer live full-time in the area, and they seem to get used to the bags in a week.

Most of the damage is done by females with fawns. "Every new generation picks out a plant we thought was foolproof," Westley says. But tulips, until they are fully open, and roses have always been favorite foods. Moreover, the deer seem to have a particular appetite for tulips just as the park is gearing up to greet visitors at its annual tulip festival, and roses move to the top of the menu just before the annual rose festival is slated to start. "Do people want to take photographs of the bags?" Westley asks rhetorically. "I don't think so."

How large and tame is Boerner's deer population? When Westley drives into the garden area at night, sometimes he has to drive right up to the deer before they move at all. At the sound of his car, they'll just look up from whatever they're munching on, as if to say, "Buddy, we're busy here."

The park comprises three square miles of deer territory and is home to at least 200 deer. Westley would like to see a population of no more than 60 or 80. But visitors like to see the deer. And even if there were fewer animals, "it only takes ten deer to cause a lot of problems," when they are coming back to eat the plants night after night.

The secret to the success of the snow fencing seems to be the size of the area that's fenced, rather than the height of the fencing that's used. Deer seem reluctant to jump inside and be enclosed in a small area.

Westley and one of his gardeners agree that, so far, at least, the deer won't touch red salvia, snapdragons, or the huge drifts of daffodils that dot the park in spring. They probably won't eat castor beans and cleomes. But it's probably not the cleomes' height that keeps the deer away. In fact, "They seem to like not having to bend over," the other gardener remarks, pointing out that the park's concrete flower planters are "table height" for the deer. "For them, the planters are just like feeding troughs," she remarks.

The problems are not confined to public parks. Deer devour flowers and vegetables, shrubs and trees, in many urban areas. More seriously, they are causing an increasing number of automobile accidents as they ghost across roads at twilight and on through the night, especially during the fall mating rut. The question is, how did deer turn from wildlife we welcomed to problems?

WHERE DID THE SURPLUS COME FROM?

In his very readable book, *Heart and Blood*, Richard Nelson begins the answer to that question with the professional hunters who stocked tables with wild venison until the beginning of the 20th century. By that time, they had made wild game so hard to find that hunting professionally was no longer profitable. Then the invention of the automobile, hand in hand with the demise of the profession of market hunting and stringent hunting regulations, brought deer back from the brink of extinction, because as people traded in their horses for automobiles, the need for hay declined. Hayfields were left to lie fallow, then natural succession yielded, eventually, forests that became prime habitat for deer.

A U.S. Fish and Wildlife Service biologist checks for deer browse at a wildlife preserve near the Minneapolis/St. Paul airport. (USFWS/J&K Hollingsworth)

Recently, though, as farmland and forest became subdivisions, suburban gardens and ornamental plants seemed logical replacements for lost cropland, at least from the deer's point of view. Deer are able to eat and digest just about anything. They have occasionally been known to eat small, herring-like alewives, and one naturalist watched a deer kill and eat a black sucker fish. In fact, Nelson writes, "these animals have become entirely too adaptable for some people's taste — dodging traffic at dusk, feasting in yards and gardens, oblivious to householders' complaints of freshly browsed roses, nipped delphiniums, and shorn lettuce. Millions of North Americans are discovering for themselves that deer can live almost anywhere." In order to prosper in cities, the only thing wild deer have to alter is their natural fear of human beings.

Nelson cites William Ishmael's study of the progression of deer into cities. Ishmael, a Wisconsin Department of Natural Resources biologist, found that first a few deer move into a new urban habitat from overcrowded wild places. For the first 20 to 40 years, the increase is gradual; each year there will be a few new deer plus the offspring of the pioneers. But finally there will be a critical mass of deer and the well-fed urban does, capable of producing two or even three fawns a year, will cause a population explosion.

Then the wholesale destruction of shrubbery begins. Before long, the deer are not as well fed, and that can cause problems: an individual deer can suffer from many illnesses, but epidemics don't usually occur unless very high deer numbers lead to poor nutrition in now-weakened individuals. Deer weakened by severe winter conditions or substandard habitat might become prey to coyotes or to dogs running loose.

But for the most part, deer are successful. Too successful. Especially where hunting is not allowed. "We have to do what predators used to do — kill them," says one of the people in charge of the Madison Arboretum. He dates the park's problems to the period when activists forced a temporary end to the use of professional marksmen who kept the herd in check.

HUNTING AS AN OPTION

In an Ohio State University Extension publication, Thomas Townsend writes, "The most cost-effective way to control severe deer damage is to remove the offending deer. This may be done through sport hunting or the use of an out-of-season kill permit." The author suggests that landowners encourage sport hunters to come onto their land by running local or regional ads, and even charging a fee to hunt.

As with geese, hunting — coupled with the use of a propane exploder between the times that hunters visit your land — will possibly keep deer away from what you don't want them to eat.

Nevertheless, Townsend warns, sport hunting is not a long-term solution, since healthy deer tend to reproduce faster than they are removed.

Ohio residents are advised to contact their county game protector or the nearest Division of Wildlife district office for information on how to go about getting a permit to shoot crop-eating deer out of season. (109) Residents of other states should contact their own natural resources department for similar information.

REMOVING TEMPTATION

Hunting, of course, is not an option for the private landowner in an urban area, even if he agrees that hunting is an acceptable answer. There are other options, some more successful than others. The other options involve turning deer away or keeping them out, but as the gardeners at the Milwaukee public gardens have found, success with many of these methods involves switching from one to another.

Wrapping tree trunks with wire mesh or plastic sleeves seems to keep bucks from rubbing bark until the trees are effectively girdled.

To keep deer from browsing trees and plants to extinction, one possibility is to

Young, supple trees are protected with plastic sleeves against bucks who might rub against them.

try to plant things that deer won't eat. It's a lot easier to find lists of things that deer like: in Vermont, the list of deer treats includes white cedar, hemlock, striped, sugar and red maple, witch hobble viburnums, dogwood, white ash, oaks, yellow and white birch, cherry, and poplars. (106) *Wildlife Habitat Management of Forestlands, Rangelands and Farmlands* lists trees that are important in Maine deer yards for winter food; the list calls arborvitae or white cedar "the best single browse species, but too palatable." (89) The New Jersey Division of Parks & Forestry, Forestry Services, makes a sheet available that lists 61 trees and shrubs that deer will severely damage occasionally and 24 trees, shrubs and plants that deer will frequently snack on. (81)

As for trees that deer will not eat, in Wisconsin, field observations showed that deer would eat certain plants only if they were starving. These included Norway (red) pine, balsam, alder, raspberry, larch and white and black spruce. (72) The New Jersey publication says that

Deer don't seem to care much for wild strawberries, but children do. Here, the authors' granddaughters raid a wild strawberry patch. (Julie Margerum-Leys)

deer rarely damage barberry, paper birch, common boxwood, Russian olive, American holly, drooping leucothoe, Colorado blue spruce and Japanese pieris. (81) Payne and Bryant list spruce, white pine, and beech as having little to no value as deer food. Though aspens are good deer food, they will grow too tall to browse in three to six years if they are not browsed heavily. Striped maple also grows out of reach of deer very quickly, while mountain maple, viburnums and willow, though good food, are browse resistant. (89)

REPELLENTS

Browse — the trees' leaves, buds and tender branches — and crops that a landowner wants for himself can be protected to some extent by repellents. Repellents vary from chemicals to organic materials like human hair or cat urine and feces. Some are more effective than others (15), and some that are successful for one landowner have not been successful in keeping deer away from the same crop in a different area. All need to be re-applied. Moreover, repellents won't work unless the deer population is relatively low, the deer have lots of alternative high-quality foods available to them, and the vegetation under attack is not something that deer prefer over most other things. If your cropland, orchard or garden is surrounded by good deer habitat, repellents might be useless. Also, the repellent must be suitable for the time of year and condition of the vegetation when damage is most likely to occur. (109)

Some of the repellents work because they taste bad to deer, but they're only effective on the surfaces on which they've been sprayed. Plant material that grows after the spray is used will not be protected. Other repellents work by making an area smell bad. However, these only cut down on browsing; they don't eliminate it. They wash off easily, and work best if they're used before deer feeding habits are established in late fall and spring. Unfortunately, one success doesn't ensure perpetual success. In the publication *Controlling Deer Damage in Maryland*, Jonathan Kays writes, "A repellent that works in one area may not work elsewhere, even if the crop and conditions are similar to the first site."(60)

Orchard owners have had some success with tankage — offal available from animal packing plants and some feed dealers. Small cloth bags sewn from scraps are filled and hung from trees. One bag has been enough to protect a small tree, but up to four are needed for larger trees. (133)

The experience with deer repellents in Indiana shows that high levels of dew reduce the effectiveness of chemical repellents to such an extent that wildlife biologists are advising that their effectiveness is limited. However, the biologists concede that repellents might be effective when coupled with other methods. (52)

Variety may be the spice of life for humans, but for wild animals, something new is something to be feared. Much of the effect of using chemical or other repellents is due to this fear. So the key, as professional gardeners have found, is to keep changing what you use: put out old rags soaked with new scents, but vary them with plastic bags or aluminum pie plates that will flutter in the wind. (109) If you're interested in more information about repellents, take a look at Table 6-1.

OTHER REPELLENT DEVICES

One alternative to repellents is noise. A loud radio might work, but it has to be moved frequently. For small properties, a much more elaborate solution involves an ultrasonic device triggered by an infrared motion detector that turns on a sound that humans can't hear. A floodlight is added to increase the startle effect. But research has shown that deer have become used to the noise in many cases. Unfortunately, dogs and cats as well as deer can hear this noise, and they might inform you that they are bothered by howling at unpleasant hours. (60)

Other devices to frighten deer as they enter an area include an automatic, bird-scaring propane exploder cannon, alternating a siren/strobe light combination with sounds of static, a steam locomotive or hard rock music, or setting off an aerial explosion of shells. Repellent sounds linked to motion detectors should be more effective than devices that operate continuously. The pamphlet that suggests these repellents points out that they are not needed during the day, when deer are usually not moving, but cautions that "use at night may disturb neighbors." (15)

If deer are ruining your fruit harvest or ornamental plants, and you have a woodlot nearby, one option might be to harvest the woodlot to provide a lot of young browse in order to draw deer away from the area where you don't want them. According to Kays, "When the creation of native habitat through good forestry is combined with an effective hunting program, deer damage can be greatly reduced at little cost." (60)

ELECTRIFYING SOLUTIONS

According to the Ohio State Unversity Extension, fencing is the most effective long-term solution, but it is also the most expensive solution. A cheaper option is an electric fence, but using one effectively means understanding a deer's behavior.

Deer travel on paths that are known to them and that they have used safely and successfully in the past to get to food. If you want deer to stop entering an area, you have to interrupt the path to the area with something new. Brush laid across the trail might be enough to break the routine and force the deer to investigate the fence you've strung for the purpose of giving him an experience that is unpleasant enough to send him elsewhere. To draw the deer's attention to the fence itself, use 6-inch by 12-inch pieces of aluminum foil smeared with a mixture of one part peanut butter, one part peanut oil or with apple mash. Fold each packet over a piece of masking tape, then tape the packets to the electric wire at five to 10 foot intervals. In trying to get to the food inside the packets, the deer will experience mild shocks and the feeding/traveling habit down that path may be broken. However, if the area you're trying to keep the deer away from contains the only food in the neighborhood, you won't have much luck. (15, 133)

A fence that is successful in keeping deer out can also be attractive.

Two strands of electrified polytape, at 18 and 30 inches above ground level, will provide control, but like so many other options, this one will work best if the deer are just starting to use the area. After three to six weeks, the deer will learn to go over or under the fence. When this starts to happen, use the peanut butter and foil sandwich routine to give yourself an additional eight to 12 weeks of effectiveness. If you're protecting a crop area, remove the tape as soon as the deer-attractive crop is harvested so the deer don't become used to seeing it and snow and ice don't damage it. (15)

Another kind of electric fence is at work at George Washington's home, Mount Vernon, where the rotational farming that Washington advocated has be re-created. The farm is down the hill from the mansion, in an isolated area surrounded by woods on three sides and the Potomac River on the fourth. It's a perfect place for deer, but deer depredation is not evident. No fence is evident, either, nor are the dogs who roam the farm at night. But there is a fence, an invisible electric fence, to hold back the dogs that hold back the deer.

Dogs must be trained to such a wire, which activates a shock collar that keeps a dog within a given area; visible indicators of the wire's location help dogs to learn.

The results of a two-year study done by Cornell University in three apple orchards showed reduced bud loss, more blooms and higher yields when dogs were used as a scare technique. (60) A Missouri study showed success in crop areas with dogs bred for herding, like Australian shepherds, heelers and border collies. The study recommended that the dogs go into the crop area at least one month before the crops become

Electrified fences can be charged with solar-powered chargers like this one.

attractive to deer so that the dogs can learn where the field boundaries are. (52) Two or more male dogs seem to work better than female dogs. Any dogs must be housed and fed inside the protected area. Though two dogs have protected 150-acre test plots in orchards, the number of dogs needed for a given acreage has not yet been proven. (15)

For more information about specific types of commercial repellents and how to apply them, suppliers of deer fencing, and a bibliography of source materials, see *Controlling Deer Damage in Maryland.*

Repelling Black Bears

As more full-time dwellings and vacation homes spread into bear territory, the search grows for ways to discourage, rather than to encourage bears, especially around houses. The Maryland Department of Natural Resources' Wildlife and Heritage Division makes a slide show available for all of the state parks in the western part of that state and the staff contacts local refuse companies about providing bear-proof trash containers. (5)

In Alaska, where the back country a bear craves is far more readily available, a University of Alaska-Anchorage student is currently studying the black bears that are increasingly sharing their native habitat with city dwellers. The student's theory is that if city dwellers understand the bears' yearly cycles, the humans will know when they are most likely to have confrontations with bears — and how to avoid the confrontations. (46)

The student studying bears lures them with moose bones, dog food, honey, bread and doughnuts. If your interest lies in avoiding bears, keep the area around your home cleared of items like these; keep garbage or refuse, barbecue grills, pet food, animal carcasses and, if necessary, bird feeders inside. Pets should be kept in protected areas. Remove potential cover by mowing around bee hives, crop storage areas and livestock holding areas. And never feed a bear, whether you're at home or at a campsite, for this teaches it that it does not have to fear or avoid humans. (99, 46)

Black bears may make pigs of themselves at bird feeders, for the seed they find there is high in the protein and fat they need. If a bear becomes a pest at your bird feeder, try bringing the feeder in every night, which is the time when a bear is most likely to be out and about. If the problem persists, remove the feeder for at least a week until the bear finds someplace else to dine. To prevent the problem, hang feeders from rope and pulley arrangements at least 12 feet above the ground and at least 8 feet from poles and trees. The feeders can be lowered for refills. Another way to prevent bear problems at feeders is to forget about feeding entirely. Instead, substitute something else to lure the birds to your property — a bird bath or dusting site or nesting box — and let the birds feed themselves elsewhere. (45)

Fences and guard dogs might also help keep bears away from places where you're not anxious to see them. (99)

Raccoons

Late every summer, without fail, raccoons would let us know when our suburban neighbors' pears were almost ripe. The pear tree was a tall, old tree and its branches hung over the fence between our properties, several of them ending 10 feet or so above the upside-down canoe that we stored on our side of the fence. One night, we would suddenly hear a thunk, the sound of an almost-ripe pear hitting the canoe's aluminum shell. There would be a pause, just long enough for one raccoon to take one bite out of one pear. Then another thunk. The tree would rain almost-complete pears for quite a while, until however many raccoons there were in the tree had decided that the fruit wasn't ripe enough after all. On one memorable night, two raccoons got to battling, possibly over territory. Whatever the subject, they chose to argue it out beneath the canoe, which proved incredibly effective as an echo chamber.

The incident seems humorous in retrospect. Far less humorous were the garbage cans we'd find upended and pulled apart — until we found a way to keep the lids on tight — and the sweet corn we lost — until we learned the secret of a successful raccoon fence.

PROBLEMS AND SOLUTIONS

Like the animals in our neighbor's tree, raccoons are almost always seen only from an hour before sunset to an hour after sunrise. In residential areas with plenty of food and cover, the raccoon density could reach one animal per 12 to 20 acres, though the average density is one per 30 to 40 acres. (8) They will eat frogs, fish, shellfish, insects, birds, nuts — especially acorns, fruits — especially apples and grapes, seeds, and vegetables — especially corn. In spring, they survive mostly on animal protein. In fall and winter, their diet is mostly plants and seeds, including ragweed. (21)

To keep raccoons away from your house, keep garbage in cans with tight-fitting lids, and move them to the curb just before pickup. To make the lids even more secure, attach them to the cans' handles with shock cords.

When raccoons can't find hollow trees for dens to nest in, they might move into those places that most resemble what they're looking for: house chimneys, attics, house walls, window wells, crawl spaces, garages, sheds, or the areas under sheds, decks or porches. (1) They're especially likely to invade if a hole or rotten wood in your roof or in a soffit or fascia allows them easy access to your attic, or a tree with a branch hanging over your roof gives them a highway to your chimney. (7) If a part of your residence has been mistaken for a den, you usually won't know about it until the babies are old enough to move around quite a bit, though they might have been born as much as a month or two earlier. (34)

Because raccoons sleep in the daytime, raccoon mothers seem to know that the quietest daytime place in your house is the area near your bedroom. This means that when raccoons are awake, you might be most likely to hear them while you're trying to sleep — and a baby raccoon's distress call sounds very similar to the crying of a human baby. Beyond the annoyance, there's a health hazard: particularly in the mid-Atlantic states, raccoons tend to be carriers of rabies. Besides, raccoons have a habit of tearing up grass in lawns as they search for grubs. (1)

When you become aware of raccoons living with you, use noise plus bright light to scare the animals away.

Or seal a pan of ammonia or mothballs in the firebox shortly before sunset. You could also put the mothballs into a stocking or prepare an ammonia-soaked rag, tie a string onto either one and lower it into the chimney; the string will allow you to remove the rag or stocking later. However, if you realize that very young raccoons are in the den, you should not use this method because the helpless babies could be overcome by the fumes. (7, 8)

After the animals leave, use a commercial cap or wire-mesh hardware cloth to cover the top of the chimney. Material as fine as window screening will clog too quickly. (8)

If a raccoon has been heard or seen in your attic, be sure you're not keeping it inside before you cover the animal's entryway. To test, plug the entrance tightly with newspapers or rags. If the plug isn't removed after two or more days, it's safe to install a permanent cover. (8)

If you're feeding suet to the neighborhood birds, hang it from tree branches too thin to support a raccoon's weight. If hummingbird feeders are attracting raccoons, remove them for a while until the pests find a new route. Put bird feeders at the top of posts smeared with the lithium grease that farmers use for their tractors. Raccoons are fastidious; they won't put up with greasy fur. (45)

If you've grown so tired of raccoons that you decide to live-trap them and move them, bait the trap with pet food or sardines. Where cats might be attracted to the trap, use marshmallows, sweet corn, fruit jam, watermelon pieces or sugared cereal. Set the trap on a picnic table or some other surface above the ground, because a trap on the ground might attract skunks. Relocate the trapped animals at least 10 miles away, where you're sure they won't cause the same problems to strangers. (8, 45)

The Illinois Department of Conservation warns residents that most live-trappings in that state require a Nuisance Animal Removal Permit. In Illinois, residents are also required to get permis-

A two-strand electric fence protects a patch of sweet corn against marauding raccoons.

sion before dumping a raccoon on someone else's property. (7) Wherever you live, it's a good idea to contact the area office of your natural resources department to see what restrictions your state might have.

PROTECTING SWEET CORN

Raccoons were particularly infuriating when — not if — they got into our urban sweet corn patch, because they would try to eat the corn when it was one or two days less than ripe, at least to our taste. In harvesting this crop that we'd been looking forward to all summer, they'd knock down a plant, strip an ear, take one bite out of it, then move on to another plant and another ear. A night's work could take care of half of our patch or more. We tried everything we could think of; we even moved our coon hound's doghouse into the middle of the patch and tied her to it. And still the raccoons came. We built a six-foot high fence, high enough to stop any animal, we figured. Then we topped the fence with an electric wire electrified with a fence charger that we attached to the fence with a buried extension cord. And still the raccoons came.

We moved to the country. It took a year before the local raccoons found our sweet corn, but only a year.

Then one afternoon, we went to a demonstration garden near the University of Wisconsin in Madison. When we drew close to their intact sweet corn patch, we nearly tripped over two parallel electric wires, four and eight inches off the ground. Just two wires; no fence. It was their raccoon preventative, one of the gardeners assured us. The theory behind it was that the marauder would put a hind paw on the hot wire that was four inches from the ground. If that didn't get him, he'd reach up with a front paw and the top wire would convince him.

So we started stringing two little wires around our corn patch every year, attaching them to the electric fence that keeps our cattle in the pasture. We could also have used a solar-powered zapper to keep the wires hot. We never lost another ear of corn, even during the summer when we trapped a record 15 raccoons as they were trying to get into our hen house.

If you are a resident of Massachusetts, the State Division of Fisheries and

Wildlife supplies trapping regulations, directions for inexpensive live traps you can build, and instructions for capturing problem raccoons, though their experts will not do the actual trapping for you. (21)

Some states allow landowners to do whatever is necessary to control raccoons on their land, while other states require permits or licenses and designate seasons. Check with a local game warden.

Rabbits

Rabbits will eat almost anything that's green, except corn, squash, cucumbers, tomatoes, peppers and potatoes. They'll also eat mushrooms, berries, fruits — especially apples, and almost all garden crops. They seem to prefer plants in the rose family: apple trees; black and red raspberries, and blueberries. In winter, when other food is covered with snow, they'll eat tree bark, sometimes with such enthusiasm that they will succeed in girdling the tree. They don't seem to bother evergreens much, but they are attracted to other young trees with smooth, thin bark and the green food material that rabbits crave just beneath the bark where it is easy to get to, though the thick, rough bark of older trees seems to discourage them. (17, 25) If you want to know why people want to repel rabbits, look no further than their diet of choice.

Since flight is a rabbit's only defense, it will seek cover in brush piles or other thick cover like thick stands of sweet clover, weed patches, berry thickets, stone piles or old car bodies and other machinery overgrown with vines and weeds. Obviously, one approach to getting rid of rabbits is to get rid of the cover that they need. (17, 25) Remember, though, that you'll also be removing habitat that is used by animals and insects you might want to encourage. (133)

Hawks and owls are among a rabbit's worst enemies. You could encourage their presence with appropriate perches. (25)

A two-foot high chicken wire fence with the bottom tight to the ground or buried several inches beneath the ground will keep rabbits out of a garden, but the mesh must be no larger than 1 square inch. To protect young trees, cylinders of 1/4-inch hardware cloth can be wrapped around their trunks. Leave an inch or two of space between the tree and the wire. Rabbits will seek out bark most often in winter when other foods are scarce, so be sure the protection is higher than the expected depth of the snow. Sometimes napthalene mothballs scattered in a flower garden will keep rabbits at bay. Dried blood may work too, but it needs to be renewed often. (25, 133) Chemical repellents like nicotine sulfate may also work. (124) Shotguns can be effective, but check with the game warden about hunting regulations and with the local police about firearms restrictions. A good dog with freedom to roam your yard is perhaps the best rabbit deterrent.

Skunks

A skunk does not initiate attacks, but reacts by spraying the animal harassing it, be it raccoon, opossum, cat, rat or family dog. (34) That spray can reach up to 15 feet. (1)

However, skunks have a redeeming social value — they eat baby mice, rats, rabbits, ground-nesting bees and wasps, shrews and moles. So, since much of their diet consists of wildlife that's even less appealing, you might want to think twice before completely ridding your property of a skunk. A better idea might be to simply keep it from building a den in the places near your nose or where another inquisitive animal is likely to seek it out — under decks, sheds, porches or houses.

The last thing you want to do is make the skunk a prisoner in its den; therefore, the first step is to be sure that it has left the premises before you seal off its home. Skunks will leave their dens at night to find food. If you sprinkle sand, dust, lime or flour at the entrance to the den, it will be easier to see the tracks that will show you which way the animal is headed. During the time when young skunks are in the den, use extra caution to be sure you're not trapping them inside. (18, 26)

If a skunk is reluctant to leave its hiding place, you can trap it in a live trap baited with sardines or canned cat food, but check first with your state natural resources department or the U.S. Fish & Wildlife Service to be sure that this is legal. After setting the trap, cover everything except the entrance with burlap so that you can move the trapped animal without being sprayed. If you don't want the skunk to stay on your property, move it at least five to 10 miles away in a sparsely populated area. Put the covered trap on the ground, open the door slowly, and let the skunk walk out. (18)

Use wire mesh to keep skunks out of openings near the ground in houses and other buildings. You'll have to use an L-shaped piece of wire, not only sealing off the hole, but the ground beneath it as well, because skunks are very efficient diggers. Moth balls or pans of ammonia might discourage skunks from returning to the area; unfortunately, these odors are also unpleasant for humans. A bright light shining underneath a deck will keep the skunk away. (34, 26)

As an added precaution, remove brush piles, stacks of lumber or other sources of shelter, and pet food, which can also be a skunk magnet.

If skunks are digging up your lawn, looking for insects, you could solve the problem by applying insecticide. (26)

Woodchucks

A woodchuck can be murder in a garden, but the secret to keeping it out is to understand what it can do.

If you have decided that a fence is your best defense, use a 2- to 3-foot-high welded mesh fence with 1- by 2-inch squares. Brace the fence securely. Because woodchucks are good climbers and good diggers, add to the fence itself an electric hot wire 5 to 6 inches off the ground and 3 to 4 inches outside the fence. To discourage digging, bury 6 to 12 inches of the wire fencing. (134)

If you're determined to rid the world of the animal, you can locate the place where the woodchuck has its burrow, then use a gas cartridge. This will release slow-burning chemicals that produce lethal amounts of carbon monoxide. This method is most effective in spring because young woodchucks haven't left the den yet. However, if the burrow is under a shed or other building, this is not a solution, since fire and toxic gas are hazards of this method. (19)

Shotguns or dogs can also be effective. See the advice under RABBITS.

Moles

Moles cause damage as they move beneath the ground looking for insects or earthworms, raising dirt mounds in the process that become bald spots in a lawn because the grass's roots have been destroyed. Direct damage to plants, on the other hand, is usually caused by mice or other rodents taking advantage of the passages created by moles. (134)

Moles are most active in the early morning and late evening on damp, cloudy days during spring and fall. They dig two types of tunnels — some in the search for food and some for shelter.

Feeding tunnels are only used once, while deep tunnels dug 6 to 20 inches below the surface are highways to feeding and living quarters and places to shelter from predators. (86)

Moles' favorite food is grubs, but removing these with insecticide might not be the answer, since moles may find other insects to eat. (86)

Small flower beds can be protected with a barrier of sheet metal or wire-mesh hardware cloth from five inches above the ground to at least a foot below ground and turned outward at a 90 degree angle for an additional 10 inches. Be sure there are no gaps in the barrier, or the moles will find them. (86)

Traps will work in an active tunnel. To determine whether the mound in the middle of your lawn is active, tramp down a short section and place a marker so that you'll remember where it is. If the mound reappears, stomp it down again. If a tunnel is raised daily, it's active and a good location for a trap. (86) Trapping is especially productive after a heavy rain because the moles will be working near the surface when the soil is moist. (134)

For more detailed information about mole traps and how to use them, see *Controlling Nuisance Non-Game Wildlife, Management Series No. 9*, from the Indiana DNR's Division of Fish and Wildlife.

Porcupines

The Cooperative Extension Service of Massachusetts publishes a series of pamphlets on problem animals. These pamphlets always include a section on the positive side of the pest. Porcupines are beneficial by virtue of being sloppy eaters, says the author of the pamphlet on porcupines. In winter, deer and snowshoe hares can make a meal from the "nip-twigs" that fall to the ground after a porcupine climbs higher in an evergreen than other animals can, nips off a succulent bough, then sometimes drops it before eating it. (12)

If you've been having problems with porcupines, you might find this small comfort. Porcupine problems generally stem from the fact that these animals like to munch on wood, and their definition of food doesn't exclude a landowner's private property, including such delicacies as ax handles, left lying by a woodpile on the theory that nobody eats an ax handle. What porcupines are really after here is the salt left behind when the sweat on the handle dried. For the same reason, porcupines have also been known to gnaw on canoe paddles. The road salt residue on car tires will sometimes attract them, too. (4, 12) To deprive a porcupine of these snacks, the obvious answer is not to leave tools or cars outside.

If your wooden porches or buildings are under attack, protect them with aluminum flashing at least 24 inches wide. Chicken wire fences topped with a single-strand electric wire an inch and a half above the fence will deter porcupines. (12)

In the woods, porcupines are partial to basswood, for this tree is high in tannic acid. (4) In New England, porcupines use wetlands and agricultural areas, but stands of mixed hemlock and hardwood provide the best food and shelter. (12) Unlike rabbits, porcupines rarely girdle trees, but they do concentrate their attention on the best fruit trees and leave other trees alone. A stand of trees near a porcupine den might show a few trees with stunted or deformed crowns, because porcupines were eating these most nutritious portions. (12)

Porcupines don't hibernate, but use winter dens in rocky hillside outcroppings, hollow trees, logs, dry culverts or deserted buildings for shelter. To porcupine-proof a woodlot, open the forest canopy and remove all slash, snags and downed logs to prevent porcupines from finding new den sites. Young trees can be sheathed in plastic tubes or wire mesh. (12)

Bats

Before trying to rid your property of bats, be aware that they eat mosquitoes and insects that eat crops, including cutworm moths, grasshoppers, corn borers, potato beetles and grain moths. Many bats eat half their weight in insects every night. In fact, it might be to your advantage to install a bat box or two for daytime shelter and possibly nesting shelter. (107)

However, an attic full of bats is another matter.

If this is your problem, try to locate the place where bats have entered your attic. Normally they'll leave their roost to feed at late dusk. It usually takes 15 to 20 minutes for all of the bats to leave. After they've left, close all the openings you've found with sheet metal or 1/4-inch mesh hardware cloth. If bats are still finding a way in after you've repeated the process several times, five to 10 pounds of mothballs might break up the roost.

Once all of the bats are gone, be sure to clean up and deodorize the area so more bats won't be attracted to it. (134)

Snakes

One of our more fastidious relatives was appalled to find snakes lounging in the sunshine on the concrete patio outside her vacation cabin. She insisted that her husband remove them. These were not poisonous snakes, but a request is a request. He removed them. Not too long after that, she found mice.

Inside. And very much alive. A lot of mice.

To her delight, the snakes did return.

If you want to keep snakes off your property, a possible solution is to remove all of the places where they might hide, like boards, flat rocks, and piles of trash, especially where tall weeds and grass are growing. Especially when the weather is cool, you might also find snakes warming themselves on open concrete surfaces.

If you find a non-poisonous snake and you want it to go, use a pair of sticks to remove it to a weedy area far from your house. Be careful, because even non-poisonous snakes will bite to defend themselves. If a snake is hiding where you can't get at it, put a wet cloth near the hole where the snake entered. Cover this with dry cloths or burlap bags. Seeking moisture and shelter, the snake will crawl under or between the cloths, and you can remove it. (134)

Invasive Plants

There are times when members of the plant kingdom can be at least as frustrating as members of the animal kingdom. Many statewide lists contain names of up to 100 of these plants.

In Milwaukee County, buckthorn, bush honeysuckle and garlic mustard have invaded the county parks to such a degree that signs urging park visitors to watch for the plants are posted in every park. Buckthorn, a Eurasian native planted in this country as early as the mid-19th century, is becoming a particular target of birders, some because it chokes out other plants, destroying

In spring, pink dame's rocket plus garden-variety orange poppies turn a hillside into something out of Monet. Nevertheless, despite its beauty, some plant biologists believe that dame's rocket is behaving like an invasive exotic.

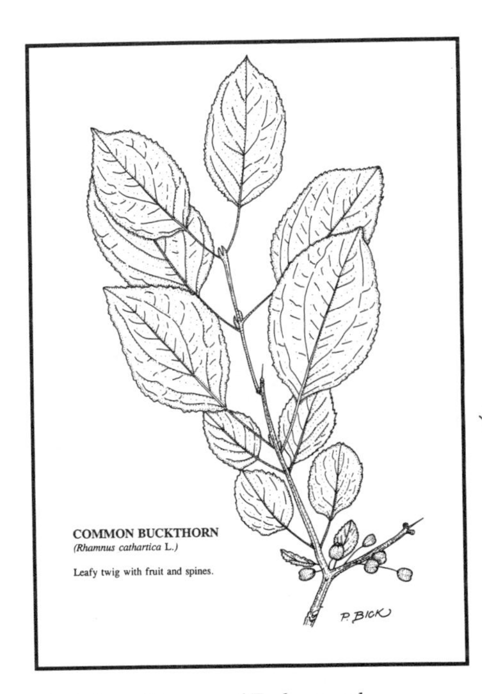

Buckthorn (Bureau of Endangered Resources, Wis. DNR)

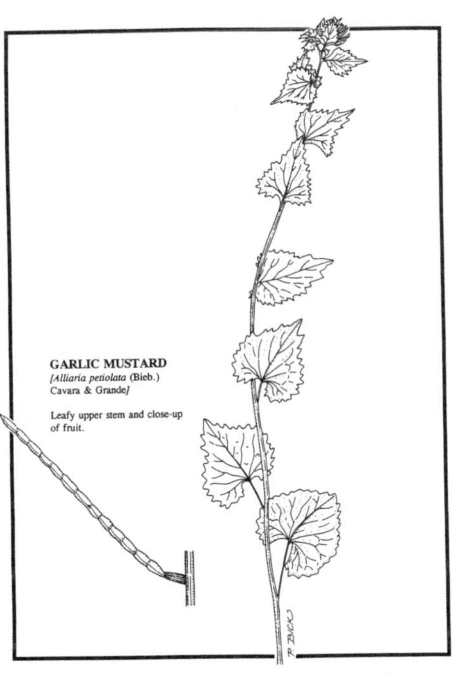

Garlic Mustard (Bureau of Endangered Resources, Wis. DNR)

diversity, others because of fears that its laxative power (its Latin name is *Rhamnus cathartica*) stresses birds.

In some cases, landowners trying to improve wildlife habitat invited exotic plants onto their land because modern experts touted them. Perhaps the most frustrating invaders fall into this category.

Some years ago, the U.S. Department of Agriculture had the perfect solution for sustaining wildlife: autumn olive. In 1972, USDA Leaflet No. 458 gushed, "Bobwhite quail, ruffed grouse, mourning doves, ring-necked pheasants, and wild turkeys find autumn olive fruit highly attractive food. But it is so tasty to songbirds that they sometimes leave very little for game birds."

The tree was admirably prolific: according to the leaflet, "Berry crops of the Cardinal strain are heavy and dependable." The seed stock came from 24 plants of a strain that yielded more than 900 pounds of berries every year.

Multiflora Rose (Bureau of Endangered Resources, Wis. DNR)

Autumn olive is in the process of taking over what should be an open field.

Plus, it was incredibly hardy: "Competition from grass, weeds, or other shrubs slows the growth of young autumn olive plants, but they overcome such competition well," the USDA assured landowners.

And therein lay the genesis of the problem. For what birds eat, they also expel. Moreover, birds have been known to fly. So lots of birds were eating lots of autumn olive fruits, flying off, and depositing the unusable portions elsewhere. The leaflet acknowledged in passing that autumn olive was native to an area from Afghanistan through the Himalayas and northern India to Korea and Japan, but apparently this rang no alarm bells for the writer.

After going on enthusiastically for seven pages about the shrub with its "reliably heavy crops of red berries," the USDA leaflet did allow itself a single, lukewarm, boxed warning on the back page: "A tendency of autumn olive to spread may keep some people from planting it. Spreading has been observed on sand blows, strip-mine soil banks, and some state game lands. In all of these places, however, spreading is desirable. There is no evidence of its spreading on croplands or well-kept pastures. It may increase on idle land as do many other shrubs such as thornapple, blackberry, sumac, and meadow rose. Although some old naturalized thickets have been reported, no places are known where autumn olive is considered a pest."

The truth dawned slowly. Spreading began to happen on some highly undesirable places, like openings that landowners were trying to maintain for the sake of wildlife. Today, "invasive" is the word that some experts link with autumn olive, although others, including the Web site maintained by Purdue University's Department of Forestry and Natural Resources, are still touting it: "Autumn olive is one of the better wildlife shrubs. Growing well on poor soils, it produces great quantities of palatable fruits with good escape cover and nesting cover."

The history of the multiflora rose is much the same, though it is still sometimes touted in manuals as an attractive "living fence." It's a fence, all right, one with barbs that catch you on the way in and on the way out. And, like autumn olive, it spreads very easily.

The USDA in its Farmers' Bulletin No. 2035, slightly revised in 1965, informed everyone, "Multiflora rose would provide

A single multiflora rosebush will grow to cover a lot of territory and provide shelter for rabbits, as the farmer whose field was being invaded discovered when he grubbed this one out.

emergency food for pheasants." The bulletin also suggested multiflora rose as a living fence to keep grazing animals away from wildlife areas.

Somewhat later, a research report issued by the Wisconsin Department of Natural Resources warned that the shrub had assumed the status of "noxious weed in numerous southeastern states . . . . several states have now banned further distribution of the species." Despite this, the report still praised multiflora rose for its wildlife value, attractive spring blossoms, hips and use as a living fence. It cited research on a group of game farm pheasants that showed they could survive on an emergency winter diet of multiflora rose hips.

By 1985, though, the Agronomy Department of the University of Wisconsin-Madison through the Cooperative Extension was issuing an alert titled, *Multiflora Rose: Friend or Foe*. The warning included four pages of advice on how to get rid of it.

Kelly Kearns, plant conservation program manager for the Wisconsin Department of Natural Resources, advises that if a plant is native to your region — the Midwest, New England, etc. — it's probably not going to be invasive on a large landscape scale. She points to staghorn sumac as an example. It does spread, but not the way an exotic might.

If a plant is not native, there's no way of telling what its history has been or might be, except by checking with your state's university extension service or natural resources department to see if someone can provide you with a list of ecologically invasive species. The fact that a commercial grower is still carrying a plant is no guarantee that it is not invasive; for instance, the Wisconsin publication on ecologically invasive plants points out, "Although [buckthorns'] aggressively invasive growth patterns have created problems in many areas, exotic buckthorns are still legally sold and planted as ornamentals." (14)

If its both non-native and touted because of it's heavy crop of fruit, ask a lot of questions about its history in your part of the country. If nobody knows the answers, try to find a different, preferably native, plant that will supply the same things to wildlife.

Multiflora rose bushes quickly invade a former pasture.

Chapter 7

LIVING WITH WILDLIFE — PROPERTY OWNERS AND THE CREATURES IN THEIR SPACE

Wildlife in a Very Small Place — Philadelphia, Pennsylvania

Mary Leupold tries to combat the wasps drawn to the hummingbird feeders on her balcony railing.

One outside balcony, 4 feet deep by 10 feet long. Almost every apartment in the Lafayette Redeemer Retirement Community has one. Mary Leupold's third-floor balcony is her garden, its railing, a place to hang hummingbird feeders.

But the space she's created for birds doesn't stop there. It extends to the small island outside the entry door three floors below her balcony, at one edge of an asphalt parking lot, plus a small space on the far side of the parking lot where she established "squatter's rights," as a gardener she says.

She's not alone in her efforts to attract birds to the sprawling red-brick complex where she lives. Some of her help comes from the tall pines and mimosas that border the parking lot and shelter the visitors to the bird feeders. More help comes from Pennypack Creek Parkway, a meandering creek plus its wooded floodplain that grazes one border of the retirement community as it journeys from beyond the northern border of Philadelphia County to the Delaware River, some 7 miles away.

Pennypack Creek Parkway provides water and cover for the wildlife that also visits the parking-lot island.

Leupold knows that she can view wildlife from her bedroom and her balcony because, in a sense, all of this habitat is hers.

BEGINNINGS

Studying and enjoying wildlife is nothing new to Leupold. "In my prime, I belonged to four garden clubs," she says. And to a nature study group that met in the evenings at a high school about a mile from her small, condominium-like rowhouse a block away from Roosevelt Boulevard, a 12-lane urban highway.

The nature study group got her interested in watching birds as well as flowers, as she took trips to such places as Brigantine, New Jersey, on the Atlantic Ocean, to see migrating waterfowl, and the flower preserve at Washington's Crossing State Park, on the Pennsylvania side of the Delaware River. "With a group like that, you gather all kinds of information," she says.

Back at home, "I wanted to try everything."

When she and her husband bought the house, the back yard consisted of a bit of soil beside a concrete pad for parking a car. They took out some of the concrete and added plants, then Leupold stretched the garden some more by adding containers of flowers, including roses, on the sidewalk. Two summers before she sold the house, she took pictures of her back yard. In those pictures, a wrought iron fence divides her garden from the concrete alleyway and the concrete back yards of her neighbors. Leupold's garden seems to be the only thing growing on the block. "They cemented, but they never parked," she recalls.

"The man I sold the house to, he cemented everything," she says, laughing.

When she moved to Lafayette Redeemer, she took along the small concrete decorations, like the sleeping cocker spaniel and the miniature Japanese lantern, along with "The Garden Prayer" on a metal stand. The only piece of garden she moved was one window box. The movers weren't happy about the weight of the soil.

In the next five years, she acquired six large white plastic tubs and several smaller pots at a discount gardening store. "I don't just plant geraniums and petunias — I plant garden plants," she says. She means hardy perennials that make it through the winter in the tubs on her balcony. Emperor tulips that "bloom like in the garden." Red lilies — "I planted them like you do anything else four years ago." They've already had to be separated and divided. One year she purchased an astilbe, though the intense heat was hard on it.

She planted two clematis, one purple, one white, in a rectangular window box that sits on top of a second window box. To arrange a trellis for the clematis was quite a project, Leupold says. The trellis is actually sunk into a layer of concrete, disguised by a thin layer of soil on top. Because that didn't leave enough soil for the plants, the second window box was added.

It was the scarlet sage in a pot beside the balcony railing that drew the first hummingbird. Leupold had tried to attract hummingbirds to her rowhouse on Ashdale Street. She even bought a book about them. But the only wildlife she got at that feeder was ants.

As soon as the hummingbird arrived at Lafayette, Leupold bought a new feeder, filled it with sugar water, and watched from the green plastic garden chair on her balcony as additional birds came by to visit, delighting her when

One of the hummingbird feeders on Leupold's balcony is shaded by the flower pot in the center of the railing. A second is near the left side of the railing.

they would fly up to reconnoiter, take a drink from the feeder, then dart away. She also added a trumpet vine to her porch garden.

Her original feeder was in the shape of a large strawberry, but it was "a dumb design for something you have to keep clean." The following year, she bought a new feeder with two holes and perches in front of them.

Three hooks hang in various places on her porch, a concession to the hot sun that beats down most of the day. She moves the feeder from place to place as the day progresses, hoping to prolong the life of the sugar water, hoping to stave off the formation of mold. Whenever the feeder is hanging at the center of the railing, it is partly sheltered. One year, a horizontal juniper in a pot shaded it; the next year, shade consisted of a pot of red petunias. A pink plastic flamingo stands guard on top of the plant stand.

One year, when September brought yellowjacket wasps, Leupold tried to battle them by replacing her big hummingbird feeder with two smaller ones, but the only birds she saw were the ones that flitted right past the new feeders. The wasps kept coming, though, despite the presence of a commercial yellowjacket trap. She tried a homemade remedy: a plant mister with salt water in it. See a yellowjacket; spray a yellowjacket; watch that wasp depart. Still new ones kept coming, so many that she had to bring all of the feeders inside.

On seeing the hummingbirds, a visitor to her apartment once asked Leupold how she does it. The visitor had hung a hummingbird feeder, to no avail. "First you have to know that they're around," she recounts that she told the visitor, then adds in an exasperated tone, "She just thinks you're gonna hang out that feeder and they're gonna come. And *then* she tells me that she has six cats and two dogs." She gestures at the patch of trees beyond the parking lot. "But look at what we have. And then there's the park and the creek . . . ."

FEEDERS ON AN ISLAND

Three flights down, beneath Leupold's balcony, a small island divides the blacktopped lane that passes

Two feeders and a birdbath draw birds to the small island.

the entry door to other parking lots. When she moved in, a hedge grew on the island, its top so carefully manicured that "it looked like you could walk on the surface." When gardeners with a different perspective were hired to tend to the landscaping of the complex, the hedges disappeared. A truck came one day, a rope was tied around the bushes, and the truck and the bushes moved on.

To the newly bare land, the landscapers added two crabapple trees, some flowering perennials, and a stand that held a birdfeeder, a plastic tube whose large holes were covered almost entirely by plastic plugs. It wasn't long before the squirrels discovered how to pull out the plugs and get to the birdseed inside. "So Mary gets to work and gets a feeder that's more suitable," she says. This one was a little plastic house, a gravity feeder that parcels out the birdseed into narrow trays through slits at its base.

Later, when Leupold realized that goldfinches and house finches were starting to hang around, she bought a second feeder for thistle seed, then a third feeder "on special." Both of these feeders are tubes, but the perches on the bargain feeder are above the holes. Leupold watched through binoculars as the two groups of finches segregated themselves. The bigger, less agile house finches stayed away from the upside-down feeder after toppling off several times, while the smaller goldfinches maneuvered to feed upside down.

By Labor Day weekend each year, Leupold can see from her balcony that the goldfinches have become so thick at the feeders that all of the perches are almost constantly occupied. When this happens, she descends and hangs another, smaller tube feeder with perches below the holes in order to accommodate additional birds. It's clear, though, which feeder is the more popular: despite the fact that it's bigger, the original feeder needs refilling more often. The goldfinches are so bold that they keep coming to the perches even as she is filling the feeder from the top, while mourning doves pick around in the bark mulch at her feet.

The house-shaped feeder draws sparrows, several kinds of sparrows that Leupold can name. There are only two birds that she is not happy to see, the brown-headed cowbird and the pigeon. She's very well versed on the life cycle of the cowbird, and speaks passionately about what happens when "the messy ones" outgrow the real nest-owner's progeny: "The little ones may perish."

The pigeons are on her bad list for a different reason: they eat the birdseed too quickly. But she thinks she found a solution. She got a plastic milk container, put gravel inside, and then when she spotted pigeons from her apartment, "Bang, bang, bang on the railing." The other birds disappeared too, but only temporarily.

The finch feeders are Leupold's project; she provides the thistle seed. The mixed birdseed for the original feeder's replacement, plus several other feeders on the complex grounds, is partly paid for by donations from a can that sits on the front desk of the complex. Another resident, a woman who lives in a ground floor apartment, reimburses her son when he buys birdseed and helps Leupold fill the feeders. Occasionally one of the gardeners will say, "I'm going to the store. Should I pick up a 50-pound sack of bird food?" His donations are not turned down.

SQUATTER'S RIGHTS AT THE SHED

When Leupold moved in, the view from her apartment took in the island, a black-topped parking lot separated into two parts by a landscaped hillside, another black-topped road, then a concrete pad, leftover from the days when a large trash bin had been installed there. "It looked like a city dump when I went out on the balcony," Leupold says. Still, there was a narrow strip of flat land between the pad and a hillside that slopes gradually, but is not something a woman in her eighties wants to maneuver.

The small patch of ground at the right front edge of the shed became a woodchuck's cafeteria. A tangle of trees beside the shed helps draw birds to the area.

Nevertheless, the year after she moved in, Leupold went out with her hand trowel, the best tool available to her, dug up the flat strip, and planted it. The little garden remained hers until the management decided to build a shed on the pad to house its snow removal equipment. The building was a little tan prefab with a hip roof, and it extended to the very edge of the slab. Two windows and a door were in the side that faced the other way, but the wall that faced Leupold's balcony was blank.

She joked with a maintenance man, "I had squatter's rights here." She pretended to be angry. The maintenance man joked back — she thought — that they'd get her a window of her own and window boxes for all three of the windows. A month later, a square was cut into the blank side "and a window appears facing my balcony." Another month and the window boxes went in, space for Leupold to plant.

The following year, she saw that a little bit of space remained on the ground where she could get at the soil at the front edge of the building to plant some annuals. But as her zinnias began to bloom, the flowers disappeared, and the cosmos foliage kept disappearing too. At first she blamed the deer, which get their water and shelter at the nearby creek and some of their food from Lafayette Redeemer's gardens. Deer can nibble a planting of chrysanthemums clean of their buds in one day. Then one morning, from her balcony, "I saw a groundhog with my naked eye." The groundhog was very fat, and it was eating the last of the zinnias. Without waiting for the elevator — it serves five floors — Leupold hiked down the steps, but the intruder was gone by the time she arrived. In the past, when she'd found three burrows in the middle of a field that borders the complex, "I went and wished them well." But not anymore.

The window boxes are another matter, though. "Groundhogs don't get up to the window boxes," she observes. In the heat of summer, she's usually outside by 6:45 a.m., carrying water in two well-rinsed laundry detergent bottles to the birdbath she bought for the island, and then to the window boxes. Each floor of the complex has its own laundry room for residents' use, but Leupold gets the water for her chores from the laundry room that's right by the entry door. She is encouraging lambs' ear to grow in the window boxes because it's a perennial, but she has bought all of the plants, including scarlet sage, geraniums and petunias, with her own money. "You can't smell the flowers when you're in the coffin," she explains.

She pays the most attention to the box facing her apartment, but likes it when other residents tell her how much they appreciate "your garden" that they can see.

But the dark moments in wildlife-watching are not confined to seeing flowers disappear. One morning, from her balcony, Leupold spied what she thought was scattered Kleenex on the blacktop near the island. When she went down to fill the feeders, she could see that the line of white was feathers. Her goldfinches? No, she decided on closer inspection, a mourning dove. "You can see where the cat dragged it," she told a visitor. "If I see that cat again, I'll hammer him to death!"

Perhaps an owl was the culprit, the visitor suggested. No, Leupold insisted. She sits on her balcony at night listening to the night sounds and sorting them out. She's never heard an owl in all her nights of listening.

A Buffet for the Birds — Minneapolis, Minnesota

It is mid-July on Kathy Sidles' long-grass prairie. A throng of bumblebees toils in a drift of light purple monarda, tiny crowns that are also known as bee balm. Bright orange butterfly weed dots the prairie, along with budding milkweed, the leaves and stems of the spiderwort that bloomed last month, and the asters and goldenrods that will show themselves in September — in all, 15 or 20 species of prairie forbs among three kinds of warm-season grasses.

A light breeze cuts the thick heat of the July afternoon, setting the prairie to trembling, just as it did in the days when this part of southeastern Minnesota was nothing but grassland. Sidles' big bluestem is only waist-high now, but late in the summer, when it's over her head, the challenge will be to keep it

Sunflowers mark the dividing line between Kathy Sidles' prairie and her neighbor John's grass.

from flopping onto the neat, evenly spaced hostas that separate her small urban property from the solid stretch of lawn that belongs to her neighbor, John.

Sidles' is not a typical re-created prairie. She doesn't have acres of ground to work with — just the larger half of a back yard in an area of fairly small one-and-a-half and two-story houses, where a neighborhood association actively works to keep the area from deteriorating. "You might want wild plants," she says, "but you also have to keep house values up." Her neighbor tells her that he likes her prairie, but she frets about what he thinks each September when the big bluestem is living up to its name.

GOING NATIVE IN THE CITY

When Sidles moved into her house, her first garden produced vegetables for her, but one summer she volunteered for a program that was measuring lead levels in the soil in this old area of town where most of the houses were built around 1910. "You can still grow a garden if you're careful about washing the vegetables, but I sort of lost my appetite," she says. "So I thought I'd grow vegetables for the butterflies."

The useable space in Sidles' backyard is about 50 feet deep and 45 feet wide. A narrow sidewalk splits the lot into uneven halves. The miniature prairie provides color and motion on the bigger half. The lawn on the smaller half is dotted with trees chosen for their value as food or cover for birds and edged by a chain link fence that also serves as a bathroom for that neighbor's dogs. A nannyberry is cradled in the elbow between the back entryway and the back wall of the house. Though the tree was planted some time ago to feed the neighborhood birds, it bears few berries, probably, Sidles speculates, because there are just too many other trees competing with it.

There are several tall blue spruce trees on the block where she lives, and the shelter these provide has yielded birds at Sidles' birdfeeders, including some red-breasted nuthatches that wintered over one year. When a neighbor gave her a blue spruce seedling that had been given away at the state fair, Sidles planted it in the middle of her backyard lawn to perpetuate the shelter if and when the big evergreens die.

When Sidles switched from vegetables to flowers, she first planted whatever she fancied. However, she got her bee balm and gray coneflower seeds from the Minnesota Native Plant Society, a statewide group that encourages gardeners to grow something other than exotics. After not many years, she noticed that her non-native flowers were dying out, but "the native stuff kept growing better and better." The mums she grew were pretty, but she never saw bees in them, so when they eventually died out, she didn't replace them.

Like the other trees in her yard, Sidles planted this nannyberry with birds in mind.

She became fascinated by the bravery of the urban natives. "Once you know native plants, you start seeing them coming up all over the city," she says. The ancestor of the columbine in her yard was a flower that grew in a crack in the pavement of the alley behind her house. "I borrowed some seeds from it one year," she recalls.

In spring, two kinds of violets bloom in her yard, Canada violets that she planted and white violets dotted with a small amount of blue that just appeared one year.

In summer, gray coneflower and oxeye daisies join the bee balm and milkweed. The big bluestem is about 3 feet tall then. Little bluestem and Indian grass grow in her prairie too. The birds, mostly juncos, appreciate her summer garden; however, "that's when you see the neighbor's cat sitting in the middle of the prairie." Once she was out watering her yard when she spotted that animal. She watered the cat, and managed to keep it away for a while. Her own cat sometimes wanders through the prairie, too, but she carefully monitors it and takes it inside when she leaves.

The birds like the native vines like hog peanut that grow in her yard. "I was pulling them up till I found out they were native plants," she says. A Peterson Field Guide helped her identify what was native and worth keeping. Before that, though, "I probably pulled out stuff I shouldn't have." The guide book informed her that almost half of her flowers were native plants that had come up by themselves. "I like to think they are still here from 1900 when this was a farm or prairie," she says.

But given the small amount of space she has, some natives are not welcome. Climbing false buckwheat, for instance, "was grabbing everything and pulling it down." Then there was the bindweed. She saw it once on a nature hike. "The vines were completely covering a dead silvery tree. It was beautiful. I came home and realized [the same plant] was choking out my garden."

ASSORTED CONTRIBUTIONS

When Sidles first moved in, the elderly couple next door had a wild border where their yard met hers. It was full of poppies and other flowers. They also had a son who was a landscaper "and the parents were too old to do anything about it." He tore out everything and replaced the patch with lawn and hostas. The next owner, John, liked the look and maintains it. However, the spiderwort that dots Sidles' prairie might have escaped before the wild border disappeared.

Other plants have equally mysterious origins. Did the yellow wood sorrel come in with the rabbit manure that she used one year when she was still growing vegetables? Did the daisy fleabane come in with the raspberries she planted? Was the row of sunflowers, as evenly spaced as the hostas that grow beside them, a gift from the birds who ate the seeds at her feeder or a neighbor's?

For the most part, though, Sidles has planted her prairie using seeds from a local supplier. The seeds go onto a bare patch of soil where a non-native plant

Sidles uses stakes to mark new seedlings so that she doesn't mistake them for weeds that have to be pulled.

has been pulled up. They surround a wooden stake so she remembers what not to pull out when the plants start to sprout. She transplanted clumps of big bluestem because these are easy to differentiate from the bluegrass that she continues to pull out whenever she finds it. She bought the big bluestem simply because "it used to grow around here." Her prairie will never have enough grass in it to be a true prairie, but at least it pays homage to what used to be native, she says.

She also encourages native weeds, because "any weed in the city is interesting," in part for its tenacity. So she has chickweed and shepherd's purse and deadly nightshade.

VALUING BEES AND OTHER RESIDENTS

Though her original intention was to plant a butterfly garden, "Instead of butterflies, I got bees," she tells friends, though she does get her share of butterflies, and these yield interesting opportunities for observations. When she spots a monarch laying eggs on one of her milkweeds, she takes note of the location and checks the progress of the larva.

Dragonflies and staghorn beetles are also drawn to her prairie, and one summer she had lightning bugs. "I have never seen them elsewhere near here," she says.

But bumblebees are clearly the majority insects. This doesn't bother Sidles at all. In fact, she seems to regard bees as some regard healthy canaries in coal mines — as indications that everything is in balance. When she discovered that a nest of bumblebees had taken up residence in a long-neglected compost pile, she simply established a new compost pile and let the bees live happily where they had homesteaded.

Maintaining a prairie in an urban setting presents its own problems. "You get bees with your butterflies, brush with your bushes, English sparrows with your native birds, disorder with the wildlife habitat." She can't burn her prairie, so every once in a while she takes her ancient lawnmower to it "to help it look a little nicer in a landscaped sort of way."

BIRDS ON THE SIDE

Busy as the back yard is, it is not the only wildlife preserve on the property. On the side of her house, in the narrow strip of land that passes beneath the dining room window, Sidles has established a dining room for birds. If she wants privacy, she draws the blinds; if she wants company, she can reach the pull cord from her dining room table.

A chokecherry and a highbush cranberry are the two pillars at the sides of the area. Between these, four feeders dangle from a bar, two for birds, two for squirrels. As soon as the chokecherry's berries turn black, the birds devour them, but the cranberries last until early spring, until there's nothing else left for the birds besides birdseed.

Beside the chokecherry grows an arrow root bush. "They're great if you like them," Sidles says. You can't kill them, but they tend to "just sort of stay down" after a snowstorm and remain sprawled to create a thicket where birds can hide. The cranberry is a "nice, upright bush," she says, demonstrating both bushes' growth patterns with enthusiastic, ballerina-like swoops of her arms.

An ear of corn dangles from each of two wrought-iron circles, food for the neighborhood squirrels. "I don't mind feeding them since life is sort of tough for them here in the city," she says.

A horizontal, open tray and a vertical, flat-faced seed feeder with two screened sides complete the quartet of feeders. An entire family of chickadees visited the vertical feeder one year. "I like to think the parents were showing the children how to use it," she says. Originally, Sidles had a conventional covered feeder, but she found that when the birds picked through the seeds to find their favorites, they scattered many seeds on the ground. They still pick around in the small, square, open feeder, but its higher sides seem to keep the seeds from falling out.

She feeds black oil sunflower seeds, but the hulls killed so much of her grass that she decided to scale back on her feeding until some of her plants grew back.

WILDLIFE PLANTINGS FRONT AND REAR

The small, shallow front yard is dominated by a red twig dogwood and a crabapple tree. The neighborhood robins appreciate the dogwood's berries, and the squirrels and birds make short work of the crabapples, but now that the trees have grown, the neighborhood mailman has had to find a different route to get from the house next door to hers. "I'm always out there trying to trim them," she says, conceding that she also contributes compost, which, of course, makes the bushes grow faster so that they have to be trimmed. "It's an endless cycle of nature I've created," she says, laughing.

At the other end of the property, Sidles' garage and the blacktopped pad in front of it form the back border of the yard. An alley serves the houses on Sidles' side of the block and the houses on the block behind it, but between alley and garage is another small strip of soil about 2-1/2 feet wide. On her 50 feet of land, Sidles has transplanted some of the prairie plants that she started from seed. These grow among the alley natives that include bouncing Bet and creeping bellflower. There's milkweed, too, but the neighborhood kids like to break off the flowers before butterflies can work them. The bees are as busy in this strip as they are in the main prairie, though they ignore the neat row of hostas behind the neighbor's garage that abuts hers. Those hostas are one more example of what she calls — with no rancor — "John's manicured look."

FINDING DELIGHT IN ALTERNATIVES

For birds that are drawn to a prairie, "space" is defined as many acres of open land. Sidles' prairie doesn't routinely draw such birds. At first, this bothered her. She wondered, "How much good am I doing here with this small little plot of land?"

But she has learned to enjoy the birds that her trees and bushes and feeders and prairie provide. She even joined Cornell University's Operation FeederWatch, spending two to 10 hours a week observing the birds at the feeders in her side yard. One year, following directions from Cornell, she put a microphone by her feeders and ran the cable to the sound card in her computer, and the computer recorded the songs. "It made me pay attention to sounds," she says. She sees cardinals and chickadees, bluejays, juncos and grackles, downy and hairy woodpeckers, crows and house finches and "a whole lot of house sparrows." Nighthawks sometimes fly over the house and she sometimes sees herons as they travel from area lake to area lake. Twice she heard a kestrel calling, then spotted it at the very top of one of the tall blue spruces on the block.

Just behind her house, across the alley, was a house whose owner had died. A wild, brushy habitat grew up. Then one day, after a storm, she heard the beautiful, flute-like song of a wood thrush. "There were dogs barking, sirens going . . . . and the thrush," she recalls. Then the next-door neighbor got enough money together to buy the house. Cleaning up the lawn was the first item on his agenda. "I suppose I won't hear the thrush again," Sidles says sadly.

She has joined a group that collects seeds on large prairies to plant in county parks that are expanding their wildflower areas. On these seed collecting expeditions, she is able to see the prairie birds she doesn't get at home. "You *can* see bobolinks. You *can* see hawks," she notes.

And if she gets many more sparrows than any other birds, she's made her peace with that, too. "I don't say, 'I get 23 times as many sparrows as everything else,'" she remarks. "I say, 'I got 23 sparrows!'"

Besides, there are always the spring and fall migrations to look forward to. That's when her feeders attract the strangers, birds just passing through, who stop by for a little fast food before heading on again. "If you want to be kind to wildlife, you become the McDonald's on the bird migration flyway," she says, adding that she wishes she could convince more of the neighbors on her block to assume the same role.

Project FeederWatch uses volunteer birdwatchers throughout the United States and Canada to monitor birds at feeders. A $15 fee covers the cost of a research kit that includes a handbook with information on feeding and identifying birds, plus FeederWatch results, a newsletter, and other items. For more information about the program, contact the Cornell Lab of Ornithology, 159 Sapsucker Woods Road, Ithaca, NY 14850, or look on the Web at http://birdsource.cornell.edu/pfw/.

Three Friends' Back Yards Equal One Bird Haven — Northeast Philadelphia

A black and white cat dangles by its front paws from a horizontal pole in front of a twin ranch a block from a busy thoroughfare in Northeast Philadelphia. In the other half of the tiny front yard, bird feeders dangle from a thick-trunked old maple. The cat is only a banner, but the bird feeders are open for business year round, tended by Angela Jacox, who was awarded a National Wildlife Federation Backyard Wildlife Habitat certificate in 1998.

Walk past the minivan that is parked in the front half of the Jacoxes' covered carport. Walk past the side door of the narrow ranch house, where the framed certificate hangs. Sit down at the outdoor table, relax, and take time to admire the view. Your ankle might brush against the sprawling blue hydrangea that hugs a post supporting the carport's roof. A hummingbird might buzz you as it ignores the ground-level pot of red impatiens and heads for the one that

Jacox and her guests can enjoy this backyard wildlife habitat with their morning coffee.

Visitors are sometimes buzzed by hummingbirds seeking nectar in a hanging pot of red impatiens.

hangs from the edge of the canopy. Another hummer might be visiting the waist-high pot of fuschia suspended from a metal hanger beyond the hydrangea in the area between the deep red hollyhock and the concrete birdbath.

At the back of the yard, tiny white butterflies sparkle in the sun as they nectar at a pink butterfly bush, one of a pair that bookends a daisy garden.

Three mourning doves waddle around the small patch of grass in the center of the backyard beneath the dogwood tree that was planted about five years after the Jacoxes moved in. Hummingbirds are often drawn to the tubular feeder that hangs from a branch in summer, birds come in fall for the dogwood's red berries, and eat mixed seed from a second, screened tube feeder in the tree year round.

Her garden reminds Jacox of Charleston, where she discovered during a tour that "unless you walk in and through, you can't see the beauty of the garden."

In this garden, perennials and a few annuals cover a three-foot wide strip around the perimeter of the 30-foot by 45-foot back yard and the even narrower strip of ground between the edge of the concrete carport pad and the chain-link fence that separates Jacox from her neighbor's carport.

FROM HEDGE TO GARDEN

Everything that is now multi-colored, multi-seasonal garden was privet hedge when Jacox and her husband bought her parents' house. The hedge benefited birds, who felt sheltered there and visited the feeder that she hung from the fence behind it. One spring, a cardinal brought her young to show them the safe source of food. But after eight or 10 years of watching her husband trim the hedge every 10 days, Jacox visited a friend who had a nice backyard perennial garden; the friend assured her that flowers were a lot less work.

As her husband went to work ripping out the hedge, Jacox planted. After two or three years of ripping out and planting, the hedge was completely gone, replaced by a wooden privacy fence along the back lot line and on the side of the house that's shared with the cat-banner neighbor.

Jacox's parents didn't mind. "My father agreed the hedges were a whole lot of work," she says.

Marian Benson, Jacox's neighbor on the other side, "was delighted when we started pulling them out," because the hedge had shaded the flowers she was trying to grow. Benson, who had lived with her husband in their twin ranch for about 19 years before the new owners moved in, was already a friend of Jacox's mother. Now she became a friend of the daughter as well. Over the years, she has shared her enthusiasm for gardening and for the birds the gardens bring in.

Benson recently started taking pictures of sparrows as they made use of the birdhouses in her yard, a frustrating exercise: she lacks a telephoto lens and the birds have a habit of popping back into the houses just before she can snap the picture.

Angela Jacox (right) and Nella Hummel pause in Jacox's carport.

When Nella Hummel, her husband and children moved into the house joined down its long side with Benson's, the most committed birdwatcher of the trio had arrived.

SHARING GARDENS

"If no one answers, come to the garden," say signs outside Benson's and Hummel's front doors. Jacox had a similar sign for a while, until it broke. They refer to their gardens as "friendship gardens," because as one neighbor has success with a flower, she passes seeds or plants along to the others. Then, if the original owner's plant fails, a new owner supplies replacements.

Years ago, when the privet hedge disappeared, Jacox grew vegetables in the plot by the back fence. When she tired of tending vegetables, she planted a mixture of seeds to supply herself with cut flowers. To her delight, the black-eyed susans in the mix came back for four or five years. Now, a daisy garden spans the area, with white shastas, black-eyed susans and other perennial daisies. Birds come to feed on the seeds she doesn't gather and pass along to friends.

A relative of Benson's supplied her with seeds for a plant with tiny, bright rose-colored flowers on thin stems, which she grew and passed along. Silene isn't perennial in Philadelphia, and the blooms won't last all summer, but Jacox has found that if she collects seeds from one year's plants and plants them at various times during that summer, she'll have flowers that bloom throughout the next summer, and seeds as well. Jacox decides when it's the right season for planting new patches by keeping track of the times when her flowers drop seeds on their own. "Mother Nature knows what she's doing," Jacox figures.

Benson and Hummel are feeding goldfinches now because of Jacox, who spotted the first bright yellow bird one summer day when it was drawn to an orange daylily. She immediately went out and bought a thistleseed feeder, then convinced Benson and Hummel to hang feeders too. To the delight of all three, the birds stayed through the winter.

Most of Hummel's backyard is taken up by an in-ground swimming pool. Hummel's first love is roses, which grow in the narrow band of soil between the concrete deck that surrounds the pool and the wooden privacy fence. But she's branching out, letting her two friends and her love of birds guide her planting choices. She bought a pyracantha specifically because its label promised "attractive fruit." She knew that the attraction was to the birds' appetites, not to human eyes. The bush is espalliered on a trellis against a gardening shed. And in the far corner of her yard, she's planted two wisteria. When these plants get old

Hummel's roses share a narrow strip of garden with flowers from her friends, a birdbath and small feeders that hang from the fence.

enough, they will climb on a trellis and shade a small table beneath it.

Small bird feeders hang along Hummel's privacy fence, offering seeds on trays. Also on the fence, 2-inch terra cotta flowerpots hang from decorative holders. Every spring, during nesting season, Hummel stuffs the little pots full of dried sphagnum moss for the birds to take. Other regular visitors include a pair of geese that take up temporary residence in the pool.

One year, a surprise appeared in the tall, narrow line of cedars edging Hummel's carport, where birds routinely shelter in winter. Robins built a nest at waist height and easily visible.

The old sycamore tree that fills Hummel's front yard offers other bird-watching opportunities. Apparently the birds feel so secure in the huge tree that her husband has been able to climb up and peer into occupied nests, checking the progress of the babies several times each spring. One year, Hummel saw fledglings half-hopping, half-staggering on the ground beneath the tree, decided they must be sick and called an expert.

The huge tree, home to feeders and bird nests, takes up all of Hummel's shallow front yard.

The expert laughed at her, she says, now laughing at herself for not knowing that birds just out of the nest are always a little unsteady on their feet.

The young pyracantha bush shares a crowded corner with a wrought iron garden bench in Hummel's yard.

OTHER ADDITIONS

Birds and Blooms is a favorite publication of all three women. The magazine's short articles, and especially its pictures, give them information and ideas. The miniature red-roofed, white-walled cottage that doubles as a birdhouse in a corner of Benson's garden was a gift from a friend whose husband refused to hang it for her. Its location, on a pole that Hummel's husband installed, is the result of a photograph in the magazine.

The magazine was also the source of information that led to Jacox's certificate from the National Wildlife Foundation. When she read about the program in 1996, she immediately thought, "I'm going to do this some day." Her garden was established, most of the feeders were in place, and birds were already regular visitors. She cut out the article, but put it

Benson's inherited birdhouse shares a corner of her garden with some of her flowers.

Despite the closeness of neighbors' houses, the butterfly bush against the privacy fence draws a cloud of white butterflies.

away for two years. Finally, her husband got busy measuring the yard for the sketch that she was required to submit, and Jacox got busy filling in the application, listing the sources of food, water and shelter for wildlife at her house. She included information about the front yard as well, "because it had some interesting things, too," she said. They include the maple for summer shelter; the bird feeders that have been hanging there for years; the pachysandra, the Jacoxes' very first planting venture, because, unlike grass, it doesn't need mowing. Beneath the maple, this ground cover offers a thick tangle of shelter for birds.

No inspection was involved before her certificate was granted. Only one warning was returned along with the certificate: "Butterfly bushes tend to overgrow." Gesturing at the two butterfly bushes at her back lot line that are squeezed into corners where small sheds meet privacy fence, Jacox says, "I can see what they mean by their growing." The plants, which are only two years old, are already as tall as the fence, and may have to be dug and moved.

Not surprisingly, Benson and Hummel are also planning to apply to the National Wildlife Foundation for Backyard Wildlife Habitat certificates.

The original motivation for Jacox's garden was to save herself and her husband work, and the promise has been kept. She moves plants around and tries new varieties, especially perennials, especially when they're on sale, but she almost never has to weed. The "friendship garden" is also a "surprise garden." Jacox jokes, "We never know what seeds we put where."

"Living in the city, you don't know what you'll get," Jacox says. But between the many trees growing on a water company easement a block from their houses and the combination of things the three friends offer, wildlife often appears. Hummel keeps a pot in her kitchen for scraps of bread that she tosses to the birds. She depends on her dog to keep away the squirrels. Benson

Jacox's garden is in the foreground, with Marian Benson's garden edged by Nella Hummel's fence.

doesn't feed bread to anyone but her family. "Squirrels dug out Marian's flower pot," Hummel clarifies.

Doves are Hummel's favorites, though Benson is not too fond of them because "they're clumsy in my rock garden and knock things over." Woodpeckers are frequent visitors, more often heard than seen as they drum on a nearby pole, sometimes knocking on the metal plate nailed to its side.

One day as she was sitting at her table and looking across the low fence that separates her house from Benson's, Jacox saw a Baltimore oriole posing on the electric line that leads from the pole to her friend's house.

"We're a bird haven in these three gardens," says Hummel.

For more information about the Backyard Wildlife Habitat program, write to National Wildlife Federation, Vienna, VA 22184-0001. Or take a look at their Web site at http://www.nwf.org. Besides the Birds and Blooms *magazine, available from Reiman Publications, P.O. Box 5294, Harlan, IA 51593, the friends recommend* The Philadelphia Garden Book, a Gardener's Guide for the Delaware Valley *by Liz Ball. Available from Cook Springs Press, 2020 Fieldstone Parkway, Franklin, Tennessee. Other states in the series include Wisconsin, Illinois, Indiana, Michigan, New Jersey, New York, Ohio and Virginia.*

Suburban Wildflower Prairies — Southeast Wisconsin

At the northern edge of the Village of Bayside, straddling the county line between Milwaukee and Ozaukee Counties, is an area of sizable houses surrounded by well-manicured lawns. One of the few exceptions is what appears, at first, to be a long, narrow, vacant lot populated by a riot of native wildflowers and a variety of trees. In spring, yellow iris soak their roots in the rainwater that runs off the street at the lot's eastern edge. In summer, purple coneflowers, yellow coneflowers, black-eyed Susans and other wildflowers fill in the areas that, in spring, appeared to be barren.

Almost 50 years ago, when Lorrie Otto, her husband, their three-year-old son and six-year old daughter moved into their house on this acre and a quarter of land, 51 Norway spruce trees and a few of the Colorado variety dotted the long lawn that buffered the house from the street. "I thought it was so charming," she says, "this little Swiss chalet with all the German Christmas trees."

Lake Michigan laps the eastern borders of the properties on the other side of that street. Most of the houses in the area were summer homes when the Ottos moved in, and each house had its own septic system. "Every other yard was a wildflower garden," Otto says, because the wildflowers covered the systems' drain fields.

But four or five years later, the area was hooked up to a municipal sewage

at all. She put up hawk silhouettes on the window, then suspended metallic ribbons outside. The ribbons rotated slowly in the breeze and cut the avian mortality somewhat, but birds were still dying. Then one hot August day, Otto heard two telltale booms against the window. Furious, she went to the garage where she was storing chicken wire fencing for use around plants, and stapled the wire over the outsides of the windows. That stopped the carnage, so the wire remains.

Dwarfed by her prairie flowers, Lorrie Otto strolls in the sun.

Otto's prairie grows all the way to the front door of her house. (Sandra Stark)

system, the other lots were divided up, summer homes gave way to permanent houses and wildflowers gave way to grass.

From the first, birds were drawn to the Ottos' property, but this turned out to be a mixed blessing. At the rear of the house, in the living room, a large picture window overlooks a wild ravine that eventually winds its way into Lake Michigan. Directly across from that window, in the dining room, two more large windows face east. When the Ottos first moved in, birds were flying through the relative wilderness of the ravine, crashing into the picture window and killing themselves, possibly in a futile attempt to fly through the house.

Otto tried putting enlarged pictures of cats on the window seat at the bottom of the picture window, but this didn't work

GENESIS OF A PRAIRIE

By the time the other houses were being converted into permanent homes, the Ottos' Norway spruces were 18 feet tall and hid the center of the yard. "All the neighborhood children played here," Otto says with satisfaction.

It was at this point that she was offered her first wild plants, some blackcap raspberries. She was delighted, because blackcaps had been a delicious part of her childhood memories. Then, one morning when she was mowing the lawn near her neighbor's yard, she noticed a purple haze on the grass. Suckers from her neighbor's plum tree were growing onto her lawn. "I rejoiced," she recalls. "There would be a plum thicket for the children." She remembered such a thicket from her father's farm. It had been a great place to play.

When someone else donated ostrich ferns, she planted them in the wet area where rainwater from the street drained onto her property.

She became the Nature Lady for her daughter's Brownie troop, which forced her to learn about the wild plants that were still growing in the steep no-man's-land of the ravine. She'd walk the Brownies around, naming the plants and telling the girls how the Indians had used some of them.

In the early 1970s, as one of the first volunteers at a newly established nature center west of Milwaukee, Otto attended her first Midwest Prairie Conference. As slides of prairie flowers were shown, she thought to herself, "When I was a little girl, that's what grew along the railroad tracks and the roadsides. I'd never thought about them until I saw them."

She started accumulating wildflowers on drives through the countryside, where roadsides were mowed and fields were planted to crops, but native wildflowers grew in the three-foot-wide stretches beneath farmers' fences. After she asked a farmer's permission, more often than not the landowner would dig up the plants for her. She'd bring them home and "plunk them into my garden," directly into the clay soil. "The things that grow out there [in the countryside] don't need fertilizer or peat moss," she reasoned.

By this time, the Norway spruce trees on her lawn were enormous and the interior of the house was dark, despite the large windows. At her husband's urging, Otto began to transform the entire lawn into garden. As she acquired more Midwestern plants from wildflower nurseries, she replaced the huge evergreens four at a time.

Besides flowers, she was moving young hardwoods in, native trees of whip size, to create woodland along with her prairie garden. She points out, though, that her creation is not a true prairie. The background is still bluegrass, and there are far too many flowers. A true prairie would consist of about 70 percent native grasses and 30 percent flowers, but the officials of her closely-settled suburb will not allow the burning that such a planting requires. "It's not all that natural," she says, "but it's native."

CREATING A SAND PRAIRIE

By the 1980s, Otto had become an acknowledged authority on wildflowers and her garden was a regular stop on charity fundraising tours. But on every tour there would be a visitor who would say to her, "I wish I could do this. But all I have is sand."

That was when Otto decided to "set up laboratories." She had the last of the really big trees cut down "on a day when both of the neighbors were gone." She arranged to have her municipality's trucks deliver dry leaves to her house instead of to the dump. When the first layer of leaves was waist deep, "the second biggest truck I've ever seen in my life" brought in two huge piles of sand. Once the sand was spread over the leaves, compressing them to a thickness of about two inches, she repeated the process. The next spring, she planted bare-rooted prairie plants into the new area and mulched them with more leaves. It was 1988, the year of a terrible drought, but the plants thrived, perhaps, Otto speculates, because the leaves held whatever water there was.

For people wanting to start their own wildflower gardens, she recommends starting with plants they'll recognize, violets, for instance, before trying more difficult species. Then they should be prepared to maintain those gardens. They should know enough about native plants to predict which ones will be aggressive in their area of the country. These plants should be removed, or maintained by bordering with other vigorous species that will confine the aggressors.

When Otto first started out, she'd always plant three of the same plant, one in each of three exposures. Then she'd watch to see "which one was the happiest."

LIVING WITH GRASS

To Otto, less grass means less pesticide, and that can only be good for the wildlife that shares the land. But if people insist on grass, they should at least hold in the water on their own land, she says. They should leave a border of natural vegetation that will soak up the rainwater that runs off the lawn, or dig out temporary ponds to hold the overflow.

particularly warm years, on particularly warm early spring days, butterflies that had wintered over would emerge into the early spring flowers.

When the March show was over, the native wildflowers would begin in the ravine, joining the twin-leaf and other wildflowers that Otto began to add to her bulb garden. She'd look forward to the yellow glow of trout lilies among the mixture of trees. Some of those trees were species more often found in northern Wisconsin, but grew in this ravine because of the cold wind that swept in from the lake and created a microclimate. A self-seeded succession of white

If a street gives you runoff, create a little pond. The street is in the foreground. (Sandra Stark)

birch and white pines was starting. There was hepatica all over the area, and so many other wildflowers that Otto began a new educational program for neighbors who wanted to troop through the ravine with her and discover the wild delights just outside their doorsteps.

In 1994, though, she stopped taking her neighbors to see the spring flowers, because there simply were not enough flowers to justify the tour.

Too many deer had discovered the ravine. Among other depredations, they would gouge out the hepatica through the snow, destroying the crowns, and eat

To her amazement, the small pond she dug as a temporary antidote to a summer of heavy rains seemed to be holding water a year later, so she was researching ways to encourage toads to live in this tiny body of water very close to a blacktopped street.

DEER PROBLEMS

For the first 15 years she lived in the house, Otto planted only spring bulbs in the shady area between her picture window and the edge of the ravine. She concentrated on flowers that bloomed by March 25, her husband's birthday. In the trout lily flowers, leaving only the low-lying leaves. They devoured the young birches and pines. The native baneberries were finally eaten by young fawns. One November afternoon, during a light snowfall, there were 14 deer just outside Otto's house. All those legs "looked like little bushes moving."

But it wasn't just the plant life that suffered. When the ravine was full of wildflowers "it was a wonderful smorgasbord for the birds," she says, "but now the cupboard is bare."

Otto decided that she and the majority of her neighbors "would finally have to go against the animal rights people." A sharpshooter is brought in each winter. Within a mile of the house, one year, he shot 52 deer.

Violets and wild strawberries seem to provide deer-resistant ground covers, and bee-balm (monarda) and other mints seem to survive too. Virginia bluebells thrive in the spring, but the leaves disappear in summer, leaving bare spots.

Otto has always kept her sand prairie fenced, but this addition has caused its own share of problems. On the day the fence went up, two thrushes flew into it and broke their wings. The first time Otto expanded the prairie into the accustomed travel lane of the deer, they nearly pushed the fence down, so now she hangs rags on a new section to warn the animals that they will have to look for a new route.

VALUING INSECTS

Otto's favorite seat is next to the living room picture window. Between that loveseat and the window, a pair of binoculars lies at the ready. Beyond the small semi-circle of spring bulbs and wildflowers, suet feeders are nailed to trees for resident woodpeckers and a wooden tray feeder at the base of the picture window keeps squirrels and chipmunks fed. In spring, migrant birds visit; rufous-sided towhees feed among the plants alongside resident sparrows. But Otto is seeing fewer and fewer birds these days. Besides the deer's appetites, she blames high wires bringing services to increasing numbers of people, increasing numbers of picture windows unprotected by chicken wire or other benign repellents, polluted water running into ditches where birds drink, and pesticides.

She talks of a neighbor who was proud of his white birches, which a commercial firm sprayed three times a year. The first spray was always done at the height of the warbler migration, the next spray when baby chickadees were just out of their nests. Otto remembers seeing

Even the driveway bricks are planted with flowers useful to wildlife. (Sandra Stark)

those young birds picking insects from the bark of the newly sprayed birches.

She talks about people who are perfectly willing to give butterflies nectar plants to sup on, but when they see the caterpillar versions of those same beautiful insects, they get out the spray cans. The larvae of frittilary butterflies love violets. "Holes in your violets should give you great joy," she says, because sooner or later butterflies will follow.

She talks about a woman she visited recently. On one side of the house was a row of bird feeders. Small, square signs were posted in the grass on the other side of the house, indicating that the lawn had just been sprayed with pesticide. She talks about bird lovers who fill their feeders in summer with hard-shelled sunflower seeds and spray their yards with insecticides just at the time when birds are seeking soft insects to feed to their young.

"Pieces aren't put together," Otto laments.

In recognition of her work educating the public about the benefits of wildflowers, Otto was inducted into the Wisconsin Conservation Hall of Fame in 1999. She is bemused by the fact that she was one of only two so honored while they were still alive. A friend joked, "They chose you because they figured you were too old to undo the good you've done." Otto was nearly 80 at the time.

Using What's Native — Ann Arbor, Michigan

The pine plantation at the front of the property made the sale. It was as if 3-1/2 acres of northern Ontario had been transplanted to a spot not quite a mile away from Ann Arbor, Michigan. These were towering pines, mostly red pines, with a few Austrian and Scotch pines, perhaps 60 years old, trunks rising many feet above the forest floor before the trees branched out. A river was the only thing missing from this miniature re-creation of the land that John and Anne Percy Knott grew to love on canoeing vacations in Canada.

The Knotts rest after a stroll through their woods. (Julie Margerum-Leys)

The Knotts' four children had grown and gone, and they were looking for a smaller house. A friend showed them the lot, one of three carved from a ten-acre parcel in an area where there are still a few farm fields and no sidewalks. The property fronts on a state-designated Natural Beauty Road, which means that Tubbs Road will remain unpaved and tree-lined, unless development and the traffic that results force a change.

Clearing land for the house cost 35 trees, but parts of two of the pines that came down serve as columns flanking the front door and wood from other trees ended up as paneling and a table.

From the first, the Knotts were determined to keep their property as natural-looking as they could. They hired a landscape architect whose work they'd admired for its wild, attractive appearance, but "I don't think our understanding of 'native' was very good," Anne Knott says now. The landscaper, it turned out, was primarily interested in show. The fringetree beside the deck at the rear of the house has showy flowers, but it is native only to areas south of theirs. The kousa dogwood is not even a native of the United States — it's an Asian transplant.

Anne Knott began to differentiate between native and non-native, and to appreciate the benefits of using only native species when she took a university course on woody plants. The course and the man who taught it were both institutions on the University of Michigan campus. Every week, the class went on a trip to see a different kind of forest. By January and final exam time, there were no leaves on the trees, so the class had to identify species by twigs and the buds on them. A major challenge for Anne Knott during this exercise was poison ivy, not because she didn't recognize it, but because she is so allergic to it.

The couple also joined the Ann Arbor chapter of Wild Ones, an organization with chapters in a number of states. Wild Ones is dedicated to "natural landscaping," the belief that native wildflowers and other native plantings are a much better idea than grass. This led the Knotts "to focus on native plant life, on saving what we have here and planting what belongs here."

Saving what is native and removing what is not has meant that John Knott spends many hours getting rid of invasive honeysuckle and buckthorn. The honeysuckle is pulled up by the roots, the buckthorn chopped out and the roots treated with Roundup. Some native plants have started to appear in the new open spaces, and other native plants have been brought in. John Knott also moves plants and trees from other parts of the property and cuttings of native species that friends have contributed from their own nearby gardens. He scattered bare-root ninebark shrubs throughout the property one spring. The name comes from the bark, which peels off in thin strips resembling the numeral nine; the shrubs came from the county conservation district.

Nowadays, before buying new trees or bushes, they consult a landscapers' bible, *Native Trees, Shrubs and Vines for Urban and Rural America*, by Gary Hightshoe. The maps in the book tell the Knotts the native ranges of what they are considering. If a tree or shrub isn't native to Ann Arbor, they don't buy it.

Hardiness is not their reason for accepting or rejecting a species. But it is the reason they have taken to purchasing their chosen trees from local suppliers. Anne Knott was born in the south and loves redbuds, which are native as far north as southern Michigan. When they started buying redbuds for the property, the trees were flowering before they were planted, before redbuds flower naturally where the Knotts live. After several of these trees didn't survive the next winter, Anne Knott realized that they were being grown in the South, and that northern-grown redbuds might have a better survival rate. Since then, the Knotts have been dedicated to buying "native things grown locally."

COPING WITH DEER

Nowadays, the Knotts don't buy smaller, cheaper trees either. But the reason for that isn't climate — it's deer.

Maybe the deer regard their property as a safe haven, the Knotts speculate. They report hearing gunshots in the area both in and out of hunting season. They've encountered people spotlighting deer on nearby roads as well. At dusk, when they return from their work at the University of Michigan, they have seen as many as 30 to 40 deer grazing in the open fields near their house.

In fact, deer have become such a problem throughout the area that the

Metropark system has proposed allowing deer hunting in three of its parks. Bowhunting has been proposed for the park near the Knotts' house. Area farmers have come to meetings on the proposal to invite people to visit their farms and look at the damage the deer have caused. Enough people are disturbed about the problem that anti-hunters and hunting supporters have been about evenly represented at the meetings.

On the Knotts' property, as many as six deer at a time have visited, nibbling the Michigan holly even though it grows against the wall of their garage. Several times, deer passed within a few feet as the Knotts were eating breakfast outside on their deck.

In the understory beneath the deciduous trees on the lower half of the property, some of the newer redbuds show evidence that they have been on the deer's menu. It's not clear why some of the trees have managed to get ahead of the deer and some haven't. In one spot at the edge of a well-defined deer trail, two redbuds stand side by side. One has been munched almost out of existence; the other is growing well.

The deer have demonstrated their preferences for what the Knotts inadvertently offer in the way of nourishment. They wait to eat the trilliums until the buds are ready to open, but the snowdrops that were transplanted from the Knotts' former house have survived. May apples, butterflyweed, twin-leaf and violets do well, and there are several thriving patches of wild strawberries. But the bloodroot seems to have disappeared. And the deer have pretty much destroyed all of the azaleas, inkberry and summersweet that the landscaper talked the Knotts into planting near the house. The non-native kousa dogwood was fairly tall when it went in, so it has survived, though the deer have thoroughly nibbled its lower leaves.

When deer seemed to be decimating the marsh marigolds, Anne Knott tried an experiment. She had noticed that wherever those bright yellow spring flowers survived in the wild, skunk cabbage seemed to be surviving too. So she gathered seeds from skunk cabbages and planted them near her remaining marsh marigolds. The next spring, a few marsh marigolds returned, but there was no skunk cabbage in evidence. But it's not clear whether the deer ate it in preference to the flowers that they were being tricked into ignoring, or whether the smelly plant simply didn't survive the move.

A fairly new patch of blueberries is covered with netting each winter, but Anne Knott wonders if they shouldn't also be using stakes to hold the netting up. As it is, the deer are pushing the net down in their efforts to get to the bushes.

A PASSION FOR FERNS

"I used to associate ferns with death and dying and decay," Anne Knott says, but now "I've gone fern crazy." The reason? "The deer don't eat them.

Though some authors say that ferns are deer food, on the Knotts' property, deer leave the ferns alone. (Julie Margerum-Leys)

Steps down a steep hillside are created by combining living and dead trees.

Paths through the Knotts' woods were mostly laid out by wandering deer, then covered with wood chips by Anne Knott. (Julie Margerum-Leys)

That's the best thing." Ladyfern and sensitive fern grow in many spots on the property. Where they don't — for instance, next to the house, and behind the house, where the septic field went in — John Knott moves them, along with sedges, which grow wild in the lower woods. A friend contributed ostrich fern, and hay-scented fern has been purchased and planted because it spreads rapidly.

Obviously well-used deer trails twist through the property. "I tried not to create new paths," Anne Knott says, explaining why she makes the best of the situation by spreading wood chips to transform narrow animal trails into wider walking paths. Some of these trails descend the steep hill behind the house to a lower, wetter woods. As trees are felled to release more desirable growth, she saves some of the logs to make steps down the slope and aid in erosion control. She props each end of a log behind a living tree that flanks the deer path, then backfills dirt. It's easier than digging the logs into the hillside, she says.

The deer continue to share most of her wood chip-lined walkways with the Knotts' guests, but in one area Anne Knott tried to curve the existing deer path around a tree in accordance with the idea, widely promoted by landscaping authorities, that a meandering path is more pleasing than a straight path. The deer will have nothing to do with her innovation.

"It's not that we want to get rid of all of the deer," John Knott stresses. Among other reasons, their granddaughter loves to see the animals.

And it's fun to watch the dark silhouettes slipping through the pine plantation in winter, Anne Knott says. But on the other hand, "When you pay for a tree, you're very invested in it."

ATTRACTING OTHER WILD VISITORS

Other wildlife, more or less welcome, flourishes on the parcel.

A wood thrush sounds reveille early each morning in spring and summer. Tanagers live in the deciduous woods. When the Knotts first moved into their new home, finding a patch of rabbit fur near the house made them realize that they'd also moved into a new lifestyle, one that they would share with resident owls. They found a screech owl's nest in a dead tree near the road, and discovered a substantial pile of great horned owl droppings beneath a massive, ancient apple tree.

A tree pathologist visiting the property insisted, "Leave something for the woodpeckers." That something was snags. They'd already discovered the benefits of dead trees, though, by watching a chickadee nest in a broken pine. However, they draw the line at diseased trees. Those are cut and the wood removed from the property. They worry, too, about the monoculture represented by the red pines that wrap around their house, for one danger in such a forest is that disease will spread rapidly once it starts.

Standing alone in an open field, the Knotts' weathered apple tree made a fine perch for a great horned owl. (Julie Margerum-Leys)

A woodchuck used to live beneath an old shed behind the house. When the shed was torn down, the woodchuck relocated to a hole beneath the big, green metal box that houses the electrical service.

The gray dogwood that flourishes despite the deer develops white berries in fall. Those berries draw cedar waxwings to the property.

Nature also found its way into their chimney, four stories above the ground. The first time the exterminator was called, a live wood duck was evicted. "It looked like a very promising cavity to her," Anne Knott explains. The chimney was capped, but not screened. The second time the exterminator was called, he found a quartet of flying squirrels, all dead. It was a hard way to discover that four varieties of squirrel, not three, exist on the property. A screen went up around the chimney opening after that episode.

The pines, though not native, provide the Knotts with birds to watch — hairy, downy and red-bellied woodpeckers, two kinds of nuthatches and pine warblers.

Even the honeysuckle has some use: before it was torn out, one thicket sheltered ruffed grouse, and robins built nests in other shrubs.

Vinca covers the front yard. John Knott planted it when the couple first moved in. It's not native, of course, but he wonders whether that's all bad. When the vinca is flowering, it's covered with bees. "The bees don't care if it's native or not," he says.

Wild Ones has local chapters in 11 states: Illinois, Iowa, Michigan, Minnesota, Missouri, New York, Ohio, Wisconsin, Kansas, Kentucky and Oklahoma.

Urban Birds and Butterflies — Milwaukee, Wisconsin

Thanks to the hospitable habitat that surrounds them, Reicherts' nest boxes are usually occupied. (Sandra Stark)

Where Appleton Avenue becomes Highway 41 on Milwaukee's northwest edge, Jim Reicherts maintains a wildlife refuge covering an acre and a half. His house is just down a low bank and across a two-lane frontage road from and well within earshot of the four-lane highway.

A Cooper's hawk shares the property, two red-tailed hawks visit regularly from their home across the highway, and a kestrel sometimes perches on the power line that parallels the frontage road. Late one winter, a pair of horned owls spent noisy nights in Reicherts' trees, and on one clamorous occasion, mated on a neighbor's television antenna.

Most of his neighbors tend lawns, some of them large by city standards. A lawn also wrapped around Reicherts' house when he bought it. A gardener by trade, Reicherts had spent a lot of years dreaming about what he would plant in his own yard, if he ever had a yard to plant. The first summer he lived in the house, he simply let the grass grow, in

Reicherts' greenhouse provides plants for his customers and his own property. (Sandra Stark)

order to see what might happen along. In September, he was surprised by a haze of purple asters growing in the grass. "It was a great way to finish off the year," he remembers.

Because he's so busy gardening for others when he should be planting at his own place, he has done very little to prepare the soil. Instead, he simply plunks new plants into the existing sod. Instead of creating formal beds, he has scattered most of his plants, giving the former lawn the look of a wilder place, where self-seeding is the rule.

Many of his plants were leftovers given to him by friends who run nurseries, but he drew the line at any wildflowers taller than 3 feet. The lot is just too small for taller, rough-looking plants, he feels.

PLANTING FOR BUTTERFLIES

Many of the plants were added specifically to draw butterflies. Purple coneflowers and red bee balm went in early on. Later, he experimented with tougher flowers like phlox, mountain bluet, and maltese cross. Joe Pye weed and non-invasive thistles grow in his yard as well, along with blue hill catnip, Missouri primrose and phlox. Because butterflies can't get inside showier double flowers, he chooses the flat-faced single varieties, but "whatever you offer, somebody is gonna go for it," he says.

In summer, there are always half a dozen monarchs and swallowtails hanging around. Reicherts sees red admirals, fritillaries, checkers, whites and sulphurs.

Mature willow and ash trees, legacies of the previous owner, provide caterpillar food. Every year, a swallowtail lays its eggs in a large willow tree near the house, then patrols its territory, aggressively chasing any other swallowtail in sight. Every summer, a red admiral "owns" the patio.

The trick, Reicherts says is to "give butterflies a way to have their whole life in your yard."

OTHER WILDLIFE/OTHER PLANTS

He's also planted with other species in mind. In an area between the driveway and the property's edge, there are cherry trees for birds and wild grape vines. But because the grapes can be aggressive, he has put them in a spot where they won't take over the entire place.

Bright red canna lilies draw hummingirds onto Jim Reicherts' patio. (Sandra Stark)

In late spring and summer, hummingbirds zero in on the red canna lillies in pots on the front porch. Each spring, pairs of hummingbirds and orioles visit the Ohio buckeye tree that towers over the house. He remembers vividly the night about 20 apricot-colored datura flowers opened simultaneously and were visited by a congregation of moths.

In the recreated wood edge on one border of the property, some of the bushes have been planted specifically to please wildlife, and some to please Reicherts. He's put in yellow dogwood for the winter interest the shrub's bark will give. A short distance away, the golden trunk of the amur cherry trees and the red dogwood's spiky limbs will provide additional interesting contrast in winter.

The clove current's small, fragrant yellow flowers add perfume to the garden in April. Serviceberry (amelanchier) provides berries for the birds in summer and mountain ash and crabapples supply fruit later in the year. Wrens, chickadees, nuthatches and warblers go after the alpine currents, and clearly value the cover and berries the adjacent shrubs provide.

Wild sweet william and dames rocket fill the spaces between the bushes. Dames rocket may soon be declared an invasive species in Wisconsin, Reicherts cautions. Despite its beautiful rose-colored flowers, it's actually a non-native mustard. However, it's great for bees and butterflies. There's also creeping charlie in this area. It's very invasive and he certainly wouldn't suggest deliberately planting it, "but bees love it," he shrugs.

In a plot next to the house, sedum spectable will bloom after the first frost, providing good nectar for migrating butterflies.

Behind the house, chickadees moved into a tree with a sawed-off top, and woodpeckers, brown creepers and nuthatches also were attracted to it. A nearby hawthorn draws the little birds, who nest in its branches, protected from predators by its thorns.

Reicherts has even transformed what others might consider a liability. He calls it his marsh, but it's actually a channel for rainwater and the place where the hose from his neighbor's sump pump drains. Here he has planted

Bird feeders are scattered throughout Jim Reicherts' property. (Sandra Stark)

Drain water from Reicherts' neighbor's sump pump has helped him create a miniature marsh.

moisture-loving trees and shrubs, and Lord Baltimore hollyhocks.

This neighbor seems to be catching on to the benefits of having something more than lawn — she's allowed Reicherts to expand his woodland onto her property.

DEER PRECAUTIONS

Most of Reicherts' clients own homes near the 90-acre Schlitz Audubon Nature Center in the northeast corner of Milwaukee County, where too many deer are a fact of life. When he moved in across town, he didn't wait to see whether deer would be a problem. Instead, he built an attractive six-foot-high fence around the vegetable garden his partner requested. Any flowering plants that deer particularly like remain in this area, too. Throughout the rest of the property, young bushes and trees are surrounded by chicken wire, protection against resident animals that include the 10-point buck that stood one morning very close to the front porch of the house.

Glory-of-the-snow and daffodils in spring survive the deer onslaught. Bees like Virginia bluebells, but deer don't, Reicherts reports with satisfaction. On the other hand, he was chagrined to discover that hydrangas are "good deer food."

Crocus planted outside the fence was intact thanks to deer repellent until they bloomed. At that point, rabbits ate them.

Besides rabbits, woodchucks and voles are resident pests. The voles, however, are kept in check by the resident hawks, plus Reicherts' black lab, who loves to hunt them.

It took about two years of planting hundreds of flowers and shrubs before Reicherts noticed an increase in the wildlife in his yard. But he also gives some credit for the increase to the previous owner, not because of what he did, but because of what he didn't do. The previous owner had a lawn, but he never used herbicides.

A deer repelling fence encloses deer attractive flowers. (Sandra Stark)

Creating a Hunter's Eden — Southwest Wisconsin

"It's deer candy," says Gary Harden, frowning, as he keeps an eye on an expanding patch of foot-high white cedars. Beyond Harden, a forester for the Wisconsin Department of Natural Resources, a tractor crawls across a former hayfield. The tractor pulls a bright yellow tree planter. Pre-emergent herbicide dribbles from the barrel mounted in front of the planter as a blade knifes a narrow opening. Then the man crouched inside the yellow box feeds another foot-high sapling into a chute, the cedar drops into the opening, and a pair of wheels at the rear of the contraption closes the earth around the little tree.

"It's deer candy," says Kurt Larson, smiling. Sitting at his kitchen table, 100 miles from the land he's bought to hunt on, he points to the box on the chart of one of the publications he has studied before choosing to have 1,000 white cedars planted. The box warns that this tree, more than most, is prone to nibbling by deer and rabbits. But what Harden and the pamphlet see as a drawback, Larson views as a benefit. His aim, after all, is to provide food as well as opportunities for cover for the deer that are somewhat scarce in the corner of the rural county where his property is located. There aren't nearly enough oaks on his land to provide food now, and new trees wouldn't produce acorns for more years than he's ready to wait. So white cedar, or arborvitae, is his tree of choice.

With proper weed control, Harden concedes, most of the cedars could grow so quickly that they are beyond the reach of browsing deer. He has directed the crew to plant the cedars far enough apart so that the strips of grass between them can be mowed. The herbicide will take care of the grass surrounding the little trees for the first year, but after that, for the next two or three years, they will have to be weeded. If Larson has only enough time for either mowing or for weeding, the weeding will be more important. Besides, Harden says, with any luck, 1,000 white cedars are more than the deer can nibble to death before the trees grow branches that are out of reach. He points out that there are places in northern Wisconsin where he has seen browse lines on white cedars, showing that if the trees get big enough, they can survive even heavy deer traffic.

But the very best deer control will be "to pick up a few bonus hunting permits," the forester says.

Which is fine with Larson, who figures he's got about nine years to improve this land. Nine years until his son is old enough to hunt along with him. His older brother's son will reach hunting age in six years, and his brother-in-law is anxious to bring his daughters to see the wildlife.

Hidden inside the box-like tree planter, one worker plants Kurt Larson's white cedars while a second worker runs the tractor that pulls the contraption.

GENESIS OF A PRIVATE PRESERVE

Some years ago, Larson was scouting before duck hunting season in what he thought was a public hunting area. Suddenly, he was confronted by an irate landowner brandishing a canoe paddle. That was when Larson decided to buy his own land and shape it for hunting.

He bought a rectangular piece of ridgetop land with a fringe of trees on its south and west borders, a woodlot totaling 19 acres. The parcel also contains about 30 acres of former cropland stretched across two hillsides. The valley where the hillsides meet is marked by a staghorn sumac thicket that leads to a small pond formed by an old erosion-control dam.

The previous owner had entered the cropland into the federal Conservation Reserve Program, where it was scheduled to stay until 2002. Larson applied to enroll his woodlot in Wisconsin's Managed Forest Lands program.

In June of 1998, Harden came out to evaluate the 19 acres of forest. He was looking at whether the woodlot met the program's minimum requirements, whether the stocking rate was adequate — whether the number of trees or the volume of harvestable wood per acre was high enough — and whether there were any domestic animals grazing on the property. For this program, the only things that would disqualify a property would be a herd that continued to graze in the woodland, or trees so widely scattered that the property was actually savanna instead of forest. The forester gathered data in sample plots in each of the five stands he mapped, took borings to discover the age of the dominant trees and figured out a site index that showed the expected quality of the stands in 50 years. He looked at the understory to see if new growth should be managed in any way, and whether the seedlings foretold a new forest like its predecessor or a woodlot that was in the process of converting to something else.

The primary purpose for the Managed Forest Lands program is to provide advice that will lead to good management practices on privately owned woodland, so that the landowner can keep timber available for one of the

The additions to Kurt Larson's land take advantage of what his scouting turned up. A key to the drawings:

T=area where turkeys were heard

D=area where evidence of deer was seen

#1=block of white cedars

#2=wildlife packet of shrubs and trees

#3=deer hunting stand

#4=block of Norway spruce trees hand-planted by Larson

o=apple tree planted by Larson

(Patricia Gilbert)

state's most important industries. Following timber-management recommendations also leads to a break on local property taxes in the Wisconsin program. But "wildlife management and timber management go hand in hand," Harden says. Good timber management will probably have a good effect, especially on game animals.

Since Larson had stated his interest in increasing the deer, turkey and grouse populations on his land when he filed the application, Harden also made recommendations for doing this. When he senses that all a landowner wants out of the program is a tax break, the forester recommends different things. For instance, instead of urging that the landowner save mast-producing trees and snags, he might recommend clear-cutting a stand.

LONG-DISTANCE PLANNING AND PLANTING

Larson's brother, a pilot, took aerial photos of the property at the time it was purchased. When he was ready to begin making improvements, Larson drew two rectangles and a triangle on a computer-generated copy of one of the photos to show Harden where to plant the cedars and a wildlife pack of 200 shrubs and 100 conifers that came from a nearby Wisconsin Department of Natural Resources nursery.

Larson's drawings were not done to scale, however, so the actual plots were smaller. The north rectangle was longer and narrower than the drawing, and Harden had the planters locate it a little closer to an existing stand of trees on a south-facing hillside. He left a travel corridor where deer could also safely sun themselves. Running his forefinger over the aerial photo, Harden indicates that when the white cedars are taller, a deer in the open area will be inconspicuous except to someone standing at either end of the lane.

The wildlife pack was planted in a triangle, the cedars in two rectangular blocks, all in the former field, where clearing existing trees was not a problem. The new blocks of trees "aren't molded into the existing forest," Larson explains, for his aim is to increase the amount of edge.

This landowner had to depend on the forester to make the final decisions about where the trees and shrubs should go. He also had to trust the expertise of the three-man crew that the forester recommended, because Larson works near Milwaukee, about 150 miles away from his Crawford County property. In fact, 70 to 80 percent of people Harden works with live in Madison, Milwaukee or suburbs of these, or in northern Illinois.

Larson studied books and pamphlets before deciding what he wanted to do and where he wanted to do it. On hikes around his property, he took note of the bedding areas that dot the tall grass in his abandoned field and the deer trails that run through his woods. In spring, he listened to turkeys gobbling and remembered where he'd heard them.

When he first bought the land, the neighboring farmer to his east was still growing corn. "Every time I sat out, I saw deer come out of there," Larson says. Now that his neighbor buys the corn he needs for his dairy and beef herds instead of growing it, Larson is looking into planting a food plot, perhaps using perennials; he has pamphlets about such a mix that he picked up at a deer and turkey hunting show he attended.

PROVIDING FOOD AND COVER

Two miles north of Larson's property is one of Wisconsin's two biggest commercial apple-growing areas. Hoping to keep more deer on his own property, Larson planted apple trees before he planted anything else. To protect his trees until they are big enough to fend for themselves against the deer, he strung small pieces of Ivory soap and hung them from the base of branches. He also wrapped two of the trunks with plastic protectors. At the end of the first winter, one of the trees seemed untouched, but there was a mouse nest at the base of the other.

The following spring, he planted 14 more apple trees, choosing a few that the Gurney Company catalog identifies as "hardy and early to bear." The majority were bought from a Michigan company that had a booth at the deer and turkey show. He paid about $6 apiece for these trees.

Harden supplied him with a written forestry plan that outlines what Larson should and must do for 25 years with the acres of forest that are enrolled in the state program. The plan breaks the 19 acres into five stands.

"I'm not looking for value. I'm looking for cover," Larson says. He's fortunate — according to the plan, Harden didn't find much immediately salable lumber in these woods. On the other hand, he did find many features that should be good for wildlife. For instance, one six-acre stand was "poorly stocked with large sawtimber sized trees," the kind of trees that tend to shade out new growth and discourage the browse that attracts deer. But "the stocking, age of trees and species composition is quite variable within the stand." Within the six acres there are open areas, some containing "dense herbaceous ground plants, while others contain patches of sugar maple seedlings and saplings." In other words, this multicultural little woods ought to be able to attract a variety of animals.

The plan notes, "Some of the largest trees are defective and have little to no commercial value, but do provide benefits to certain wildlife species." It directs Larson to "girdle some of the defective larger trees (primarily the cottonwood and aspen) to create snags to benefit woodpeckers and other insect feeding bird species and cavity dwelling wildlife species." Among the pamphlets Harden sent along with the written plan was one on snags. Girdling to kill trees was not news to Larson, but leaving those trees standing was. "I thought you just cut them down," he says. But after reading the pamphlet, saving snags made perfect sense to him.

Harden describes Larson's cottonwoods as "wolfie," because of their wide-spreading canopies. If the wolf trees were oaks, he would never advise girdling, because living oaks add food. But this type of tree adds nothing while it is alive; in fact, by shading out pole-sized maples and other more valuable hardwood trees, it detracts. Yet if such a large tree were simply cut down, the forester points out, it might fall on and destroy some of its small neighbors.

On the day that the white cedars were planted, Harden also spent time in the woods, marking trees for girdling. Not all of the trees; he didn't have time for that. But enough so that Larson would understand what he should do, enough "to point him in the right direction."

In another stand, leaving "some dominant mast-producing trees, (white and red oak, hickory)" is one of the things that Larson is required to do. The plan also suggests, "Leave a couple of larger trees with cavities per acre, if available, along with any dead snags and scattered cull oaks with large crowns to provide acorns and roosting and nesting sites for wildlife."

As for the small pond with its thicket of staghorn sumac to the east and hickory and elm seedlings scattered throughout a dense field of herbaceous cover, mostly goldenrod, to the west, Harden noted that the pond "provides drinking water for a variety of mammal and bird species, as evidenced by the abundant tracks along its edge, as well as a breeding area for amphibians and insects." He recommended hand planting "clumps of white pine, red pine, white cedar and/or Norway or white spruce and shrubs beneficial to wildlife. This will improve species diversity and should enhance cover and food availability for some wildlife species and provide bedding areas."

Harden sees no need to spend time cutting the sumac. It's doing what Larson wants it to do: it provides food for wildlife. Why tear it out and plant something else that's going to accomplish the same thing, Harden asks.

One purpose of the Managed Forest Lands program is to give some tax relief for the owner of woodland whose products are never going to bring in as much money as the products of agricultural land. Larson will get a tax break on the 19 acres enrolled in the program. If he had applied in time, another program might have brought in another government agency to share the cost of buying and planting his white cedars.

"We want people to buy into the idea of doing what we're doing and be happy with the results," Harden says. Every year that he is in the program, Larson is entitled to a maximum of 24 hours of Harden's time. In addition to giving advice on site, the forester will help set up timber sales and suggest other contacts.

Timber harvests must be done if they are needed within the time period of the contract the landowner signs. And any erosion problems must be corrected. Otherwise, as Larson's choice of white cedar illustrates, this program is landowner-driven.

FUTURE EFFECTS

Will this property make a difference? Probably not to the total wildlife population of Crawford County, Harden says, but to this little piece of land? His eyes sweep the block of newly planted cedars. Not next year, but in a few years, he says, there will be deer trails visible through the stand. And buck rubs. Probably the bucks will find the tallest, straightest cedar of all to rub their antlers against. He smiles. And after that, nests will appear among the trees in the spots where deer have slept. Maybe as people drive the track that runs alongside the grove, the deer will hide in it. And the songbirds will find it. And the mourning doves.

He recalls walking through a newly mowed border around a soybean field on another property after a farmer had made just one pass with a brush hog mower. A turkey hen and a poult had already discovered the open area. If he does drastically different things and creates habitat that wasn't there before, Harden says, a landowner can expect results, assuming, of course, that the wildlife is in the area to begin with.

For Larson's property, it's a tradeoff, of course. The meadowlarks and other ground-nesting birds will have lost some of their territory. Harden's eyes sweep a neighbor's adjacent hayfield: he's looking at a lot of space for meadowlarks.

A Hunter/Landowner's Dilemma — Southwest Wisconsin

"Hjortdal" reads the rustic sign at the dead end of a township road in southwest Wisconsin. That's Norwegian for Deer Valley, the bowl-shaped valley and surrounding ridges that belong to Bud Jordahl, an avid deer hunter and retired wildlife biologist, and to whoever comes after him.

The name was an optimistic choice 31 years ago when there were seven deer per square mile in the Richland County area. On Jordahl's land, in the biggest of what had once been a farmer's crop fields, white pines self-seeded and prospered.

But area deer prospered too, and hunting pressure simply couldn't keep up with the population explosion, especially as more and more private land in the area became off-limits to hunters. In recent years, a thousand-dollar patch of Christmas trees that Jordahl and his daughter planted — but failed to shield — was almost completely wiped out.

In 1994, there were an estimated 36 deer per square mile in the area after hunting season ended. A special early doe-hunting season in 1995 targeted those areas in the state where deer were judged to be out of control; Richland County was one of those areas.

In the old days, Jordahl told a Madison newspaper, "You had to hunt, and that means understand the animals, their movements, their trails and where you should locate your stand." But with the deer population too high, "It's not a deer hunt anymore. Instead, it's a deer kill."

Nowadays, Jordahl shares a paradoxical position with many other hunter/landowners: the species of trees he plants on his land are still chosen to encourage deer, but he's been so successful that he must wage a constant battle with them. "Anything you plant, assume the deer will wreck it," he observes, laughing.

WORKING WITH WHAT'S NATURAL

Jordahl believes in working with what is natural to an area, which in the case of his property means that he puts special emphasis on oaks. Oak savannas — stands of oak trees surrounded by grassland kept open by grazing or periodic fires — were part of the original scenery here.

But "managing natural resources is an enormously complex task for both private landowners and public agencies," Jordahl wrote in a retrospective article for the newsletter of the University of Wisconsin School of Natural Resources. "To manage land wisely requires knowledge from a myriad of disciplines. The trick is to know the questions to ask, where to elicit answers and then to synthesize and interpret the results."

In choosing trees, he cautions, it's important to consider "what is state-of-the-art in terms of disease." In Wisconsin, there is more risk of oak blight in red and black oaks, so the new seedlings he had planted on his property one spring were white oaks.

Because of the deer, each seedling was protected by a plastic mesh sleeve held in place by a bamboo pole. Jordahl prefers the mesh to solid plastic tubes, which he's heard create microclimates that hamper the seedlings' ability to harden off. Besides, ultraviolet rays cause the mesh to deteriorate in four or five years, when it's no longer needed. Too, the mesh sleeves cost one-third as much as the solid tubes in the forestry suppliers catalog he orders from, and take a lot less staking to hold in place.

Each seedling would get one more piece of deer protection as it matured: in about three years, when the oak had grown taller than the sleeve, but still short enough for a deer to nibble, a piece of waxed paper would be formed into a cone and stapled over the terminal bud. This treatment would be repeated until the leader grew so high that even a deer standing on snow couldn't nip it off.

OPENINGS AND OAKS

Practicing timber-stand improvement, Jordahl, and the commercial logger who worked for him, first created openings by cutting brush, much of it bitternut hickory and ironwood, whose dense, low shade keeps more valuable seedlings from growing.

Oak seedlings in a newly created opening are protected with plastic mesh sleeves. (Bud Jordahl)

In these openings, where 40 percent of available sunlight hits the ground, the oak seedlings were planted in clumps of 10 to 20 trees, set about 3 to 4 feet apart. This was closer spacing than the 6 to 8 feet that foresters recommend, but Jordahl wanted to get the edge on faster-growing, self-seeded maples, almost as much of an enemy of his oaks as the deer. In about 20 years, if all of the oaks in a clump have survived, they will be thinned.

Jordahl instructed the logger to leave the tops of trees where they fell. In one of the new openings, tree-tops formed a thicket that made a complete ring around an area where acorns had already sprouted, and the ground was about as thick with self-seeded new oaks as adjacent areas were with mesh-sheltered seedlings. When he was a graduate student studying the effects of deer browsing on forests, Jordahl found new seedlings of a variety of trees only beneath fallen trees whose thick tops had kept the deer from nibbling the seedlings. For his current project, he is hopeful that this thick ring of branches will also hold the deer at bay, at least until the little oaks grow tall enough to escape them.

Maintaining openings, shielding oak seedlings from deer — it sounds like a lot of trouble. But any oak tree is "a living food patch," Jordahl says, that will provide food for wildlife, including deer, for up to 200 years. And mature trees that don't live that long will make valuable timber.

OBTAINING WHAT YOU WANT

The first question any landowner has to ask, Jordahl says, is "What do you want?" because the landscape you create depends on the answer you give.

The enclosed front porch on the front of his old farmhouse looks out across a grassy area to a steep hillside. Everything is a trade-off, he says. For him, the view is crucial. On that hillside he's traded some cover for a long, clear view of turkeys and deer and other wildlife descending to forage. It means maintenance: the swath must be kept clear of brush, especially autumn olive, planted some years ago as part of a project to bring grouse to the neighborhood, before anyone guessed how invasive the shrub could be. But for his pains, Jordahl gets natural shows, like the one on a winter day involving one wild fruit tree and five greedy grouse — sometimes known as partridge.

He added a feeder to the area, a black plastic barrel on legs that replenishes the grain in a hanging dish whenever an animal feeds on it and sets the dish to swaying. The grain is corn and the animals that he is trying to attract are deer, in contrast to the extensive anti-deer effort on the hillside just above the feeder. A contrast, too, to the young native red cedars planted close by the feeder. Those cedars are naked at the waist, representing the area above the snow that was low enough for the deer to reach. Belatedly, Jordahl put a screen around each of the cedars, then planted a young

Autumn olive intrudes on land that Jordahl prefers to keep clear.

spruce near each of the cedars because deer hate spruce.

He wants bobolinks and meadowlarks, so he planted a prairie in a 7-acre open space at one side of the house, mowing twice during the first summer, then spraying Roundup in the fall, then, finally, planting the following spring. Mowing and clipping the new growth, then, possibly, controlled burns, would give him his prairie, though the bobolinks and meadowlarks would not appear if they were attracted, instead, to the many vacant properties nearby. He wanted that prairie, though it meant resisting the siren song of a forester who might offer cost-sharing to transform the open field into a forest of black walnut trees.

Jordahl demonstrates a demand feeder. When a deer moves the small plate at the bottom of the feeder, corn drops in from the bin above it.

Uphill from this space, another, smaller prairie began to sprout two years earlier. It's in an opening where an old township road crests, then meanders downward again. Jordahl's son was married in the clearing. In preparation for the wedding, the father of the groom sprayed Roundup on a 20 foot by 20 foot space. When the greenery died back, the groom spaded up the square. After the marriage ceremony, guests and the wedding party stood at the edges of the bare ground and tossed prairie seeds instead of rice. The seeds had been custom-mixed at by an area supplier.

The result is a mixture of bluestem and sideoats gramma and prairie flowers from the wedding mix, plus other forbs from the property, like St. John's wort and mullein, that have crept in. Whether the wedding planting will creep outward to the rest of the opening without soil preparation is an open question.

Jordahl created a swimming pond close to the house, and he wants the spring-fed pond water to remain clear. One of a number of water diversions on the property accomplishes this for him; it's a low, grassy berm and waterway that channels rainwater around the pond, not into it. But he also wanted a bur oak in front of the house — something of a mistake, it turned out. "Those acorns aren't any fun to walk on barefoot," he says ruefully, noting that the oak stands in the middle of the natural path from house to pond. Besides, he's going to have to prune the tree to direct the branches away from the house, to keep its limbs from threatening the building and his descendants in the 200 years he expects the tree to live.

Jordahl also wanted a place where hunters and unarmed wildlife observers could remain hidden from the animals they were trying to observe. However, when the flying squirrels on his property equated the blind he built with food, he consulted Scott Craven, a Wisconsin

Bright flowers dot Bud Jordahl's son's wedding prairie.

Covered with wire mesh, this blind no longer serves as snack food for flying squirrels.

wildlife biologist whose specialty is problem animals. On Craven's recommendation, Jordahl covered the surface of the blind with chicken wire to make it inaccessible to his unwelcome visitors.

WAITING AND PLANNING

It took Jordahl about two years of hunting and hiking to learn where he wanted to make trails. Using topographical maps, he laid out roadways that mostly parallel the contours of the land, in order to keep the roads as level as possible and to avoid as much erosion as possible. Diversionary bumps that run across the roads, constructions that people in his area call thank-you-ma'ams, channel rainwater across the trails and down the sides. Where rainwater has been allowed to run straight down the trails, the erosion is evident. Spotted along the trails are stumps that look like the rooted remains of trees that grew there. They're not, Jordahl points out, tipping one up. But they make good seats for weary hikers or hunters waiting for prey.

It's especially important to plan carefully before building new trails, Jordahl says. "A bulldozer leaves quite a scar," he explains, pointing to one such scar, caused when a commercial logger created a skid road while the landowner was working at his office, 90 miles away. After 13 years, the path is still visible. Any future skid roads will be agreed on ahead of time by logger and landowner; Jordahl has that in writing.

Shade-tolerant grouse food — white clover — was planted along unwanted former farm roads. It's mowed once a year. Birdsfoot trefoil might have been a better choice, except that it requires full sunlight to thrive. To do that, far too many trees would have to be cleared.

SEEING RESULTS

Wildlife are reaping the benefits of what Jordahl has brought to the land and what he has let alone.

The first time commercial loggers came in, they cut overmature trees to open up the forest canopy, let in sunlight, and allow new growth on the forest floor. Thirteen years later, the loggers came again and thinned some pines to get wildlife-sheltering brush growing on the needle-strewn, barren forest floor. They also opened up other sections of forest canopy. For the benefit of the grouse, the tops of the trees were stacked and left behind for shelter. On some of his land, "You can go from one stack to the next and flush grouse from every one," Jordahl says. One stand of mature aspen, or what everyone in the area calls "popple," has been clearcut and a shed made from the wood. Aspen is famous for regenerating, providing a continuous banquet of young tree buds for the grouse. Other clearcuts are planned. Jordahl and the others who hunt his land with him take their grouse census during the spring turkey hunt. The number of places where they hear drumming in spring tells them how successful their fall hunt is likely to be.

CREATING SNAGS

In addition to flourishing oaks, Jordahl admires dead trees, particularly elms, for the story they can tell about the wildlife living in them. "There's great beauty in a snag," he says, stopping to point out a bleached tree-trunk where a pileated woodpecker has drilled a spiraling line of foot-long holes. One winter, the huge bird produced a three-foot high pile of sawdust at the base of this snag, providing clean bedding for the wood duck boxes that Jordahl sets near the several small ponds on the property. He considers snags so useful, for squirrels and hawks as well as woodpeckers, that when a stand of trees is improved by thinning, he instructs commercial loggers to create snags by girdling larger trees that are not useful commercially, making two cuts in the bark an inch apart, all the way around each tree. The cuts are 18 inches to two feet above the ground, deep enough to interfere with the cambium layer. Trees girdled in early spring are usually dead by summer.

Perhaps because Jordahl's aim is to create or maintain only whatever originally existed on his land, the only completely new species he's seen is the red squirrel. Years ago, when he encountered red squirrels in northern Wisconsin, his wife urged him to transplant some to his property. That was not something he could agree to; however, as time brought more evergreens to those areas of his property where a farmer once planted crops, the small animals have started to turn up.

A NEIGHBOR'S MIXED MESSAGE

Jordahl began his career as a government wildlife biologist, after completing a thesis on the impact of whitetail deer on forest degeneration in northern Wisconsin. Then he went on to become a member of Wisconsin's Natural Resources Board and its chairman for two years, an advisor to Gov. Gaylord Nelson, a professor of urban and regional planning at the University of Wisconsin and a regional planning specialist with the University of Wisconsin Cooperative Extension.

Over the years, he has learned not to parade his expertise, but when he paid a neighborly visit to the owner of the new log house down the road, he did make a few suggestions. The newcomer was anxious to attract wildlife to his 8 acres of land, though he wasn't going to hold off on changing things in order to he see what might be there already.

He knew he wanted deer, but Jordahl suggested that instead of planting bushes against the house, as planned, his neighbor should establish a feeding station where it could be seen from inside the house. The feeding station could be surrounded by shrubs, plus a salt block and floodlight, if he didn't mind adding something artificial. Prune the shrubs and they'd draw wild turkey and bobwhite quail in winter.

The township road passes quite close to the log house. Surely the newcomer wanted a screen to keep drivers from peering into his windows, Jordahl said. The newcomer had thought of that: he was about to order hybrid poplars from a nursery. Why not dig up small red cedars from your own property, Jordahl asked. "That's your nursery out there." In a few years, sitting on the porch of the house and looking at those poplars would be like looking at a row of gigantic fenceposts, whereas the red cedars could be pruned so that they would provide both a screen from the road and a view across the road to the adjacent hillside.

One of the features of the parcel that draws wildlife is a small creek that winds parallel to the town road. When the newcomer stated his intention of "cleaning up" the creek bed, Jordahl cautioned against it. Birds shelter in the brush there, Baltimore orioles, cardinals and others. It's part of a wildlife corridor, that wild animals and birds use to get safely from one feeding area to another. The area is full of farms with fields plowed border to border; it has very few wildlife corridors.

The newcomer made no commitment, but the next time Jordahl walked that way, there was already brush cut and stacked at the edge of the property.

SECURING THE FUTURE

The trend in his area, as in so many others, is for land to be broken up into smaller and smaller pieces and sold as vacation property to people who insist on bringing the city to the country. Jordahl is trying to keep this from happening on his piece of land by giving the development rights to a land trust. This will lower the resale value, but the tradeoff is a guarantee of well-managed, wildlife-rich land for Jordahl's heirs, plus a property tax break for him.

As part of the process, he has filed a management plan with the land trust, telling what he has already done, and what ought to be done in the next 50 years. The general goals were set with an eye to benefiting wildlife: establishing stands of trees of mixed ages; maintaining and creating openings; maintaining small stands of aspen for ruffed grouse, and planting "woody shrub strips and clumps along and within the forest land for wildlife habitat."

The plan gives a nod to his four-footed nemesis: "Natural regeneration of the forest, and especially the forest species which the present owner desires, will in no small measure be dependent upon populations of deer." But it follows this with a recommendation: "The present owner encourages deer hunting; future owners should do likewise in the interest of good forest management and the perpetuation of a great Wisconsin tradition, deer hunting."

A publication of the University of Wisconsin Cooperative Extension Service recommends using thank-you-ma'ams on roads and trails whenever the terrain is steep. It suggests installing these low cross-ridges at a 30-degree angle across the road, sloping them down 2 to 3 degrees.

Land can be put into the care of a private land trust in several different ways. Besides assigning development rights, as Jordahl has done, land can be donated or sold at a bargain price to the trust. A trust might pay full market value for a particularly significant piece of land. In general, a land trust is interested in preserving natural areas such as oak savannas, wetlands, recreational lands such as trail systems and public hunting and fishing areas, and scenic views or unique natural areas such as remnant prairies.

Appendix A:

A SAMPLE BIRD AND BUTTERFLY GARDEN

Working with Tables 4-1 and 4-3, Lisa Ashley of Read's Creek Nursery, Readstown, Wis., planned a bird and butterfly garden for a 15- by 40-foot area at our house, which is in hardiness zone 4. She rejected several of what seem the most wildlife-attractive choices on the tables if her customers' experience with them has not been good.

Before drawing a plan, the first thing Ashley always does is to consider the place from which the gardener will most often be looking at the garden. She also considers the height of any windows. Then she takes into account what the gardener will want to be doing. What I wanted was to be able to watch wildlife from the comfort of the rectangular area (currently a cement pad) at the rear of the garden. In addition to the planted attractions, Ashley put in a birdbath. So that I could fill it without trampling plants, she added a flagstone path.

Three features were already in place before Ashley began planning this garden, an old rugosa rose bush and a clump of peonies on the right side and a metal pole on the left. The rose bush is a convenient perch for birds in summer, the peonies attract butterflies, and the pole, a mysterious leftover from previous owners, can support a climber like morning glory, if the 5 feet closest to the ground are enclosed in wire mesh to give the climber something to start out on, Ashley says. Forget that noted hummingbird attractant, trumpet vine, she

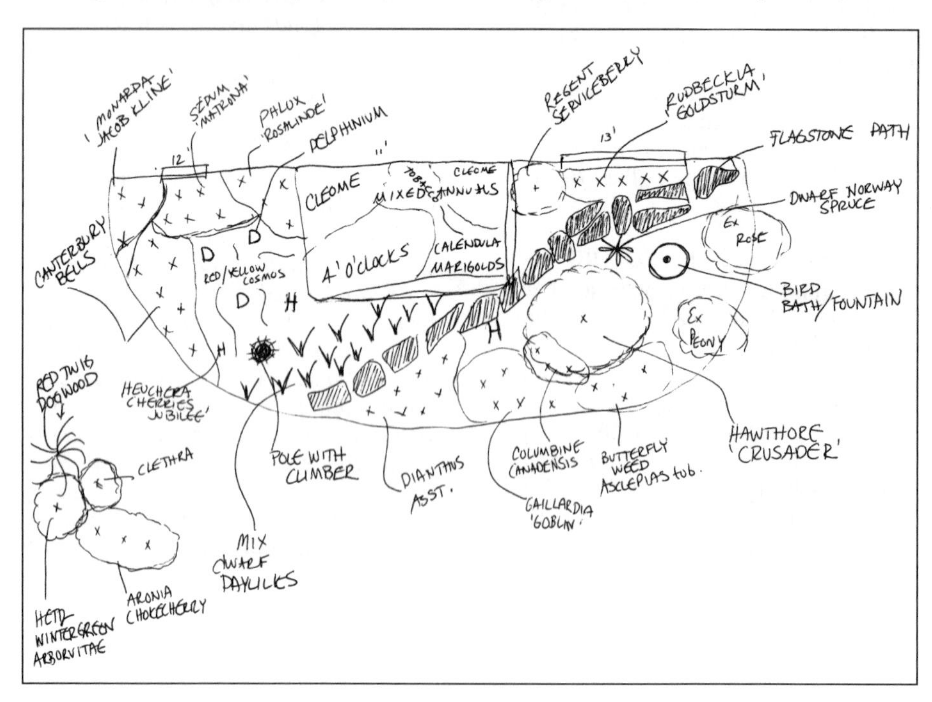

advises — this pole is far too short to accommodate it.

Trees, either dwarf or those able to be severely pruned, were planted as a place from which to hang feeders, to add summer shade and vertical interest, and to give the birds somewhere to perch while they wait their turn at the birdbath or feeders. The red twig dogwood can be pruned to the ground in March and will sprout again and bloom later in the spring. The variety Ashley specified spreads far less readily than the red twig dogwood that grows wild along roadsides in this area.

As its name implies, a hawthorn comes equipped with thorns, but because this tree will be in an area where people will be coming and going, she chose a variety with spurs that are just long enough to discourage cats from climbing to get to the birds that are sheltering in its branches.

Ashley's practice is to put in enough plants so the garden will look full in two or three years, in order to avoid excessive dividing and moving. She advises using a minimum of mulch in the interim, because the perennials should be spreading. Instead, she suggests planting annuals in the empty spaces to shade the ground and keep the weeds down. The names of the annuals she suggested for my garden are in the rectangle.

When she plans a perennial garden, Ashley arranges the flowers by time of bloom and color. She will think about what will not look good together before thinking about what will co-exist beautifully. A lot of annuals bloom in a different color range than perennials, she says, and the colors are often hotter than the color of perennials. For instance, one year she planted tithonia, or Mexican sunflower, to fill in space around some phlox. As soon as it bloomed, the bright red-orange annual became the dominant feature. It clashed with the phlox, and it bloomed a very long time. She vows she'll never plant tithonia again, except in a big swath along a woods, by itself against a lot of green.

Heuchera, or coral bells, are particular favorites of hummingbirds. The variety that she specified here has an attractive purple leaf and a flower that is clear red rather than the salmon of many varieties.

Next spring, she has been promised a heat-resistant delphinium which is probably what I will plant. The phlox she specified is mildew-resistant.

As for the monarda, Ashley chose a variety that does not spread as fast as some. Then she tucked it into a corner, next to the lawn, which is mowed regularly. A path from the door beside it to the lawn (not shown on the plan) should also keep it in check, and for added insurance, I can sink plastic edging.

Miniature daylilies are Ashley's choice for a garden plot with limited borders. The old-fashioned orange daylilies shown here need to be planted where there's lots of room to spread.

Appendix B:

NATIVE PLANT NURSERIES

Note: Wildlife biologists are becoming increasingly concerned by the spreading of invasive exotic plants. They caution that you should be sure to buy plants that are native to your region of the country. Some would like you to restrict your choices to an even more limited area. The fact that you can buy a particular plant from one of these nurseries does not necessarily mean that it is native to your area.

ILLINOIS

Bluestem Prairie Nursery
RR 2
Hillsboro, IL 62049
Ph: (217) 532-6344
Services-. Plants, seeds, consulting

Country Road Greenhouses, Inc.
19561 Twombly
Rochelle, IL 61068
Ph: (815) 384-3311
Fax: (815) 384-5015
Services: Wholesale for plants only

Enders Greenhouse
104 Enders Drive
Cherry Valley, EL 61016
Ph: (815) 332-5255
Fax: (815) 968-2941
Services: Plants, seeds, site evaluations,consulting

Henry Fromm Fromm-Huff Farm
10998 Salisbury Road
Pleasant Plains, IL 62677
Ph: (217) 626-1583 or 626-1690
Services: Wholesale for plants, retail in quantities of 100 or more

Genesis Nursery
23200 Hurd Rd.
Tampico, IL 61283
Ph: (815)438-2220
Fax: (815)438-2222
Services: Seeds, plants and consulting

Jet Hall
R.R. 1, Box 81
Walnut, IL 61376-9717
Ph: (815)379-2629 (leave message)
Services: Seeds

Earthskin NURSERY
9331 N CR 3800E
Mason City, IL 62664
Ph: (217) 482-3524
Services: Consulting, local ecotype seeds and wholesale and retail seedlings of prairie species of the Grand Prairie.

The Prairie Patch
R.R. 1, Box 41
Niantic, IL 62551
Ph: (217) 668-2409
Services: Plants, seed, consulting, custom planting

HE Nursery
1200 Old Rt. 66 North
Litchfield, IL 62056
Ph: (217) 324-4218 (ask for Henry)
1-800-761-6191
Fax: (217) 324-5756
Services: Wholesale and retail container herbaceous plants and shrubs, including many color forms and ecological selections from native populations, consulting.

Native Plant Materials
Dr. Peter Schramm
766 Bateman St.
Galesburg, IL 61491
Ph: (309)343-2608
Services: Seeds, consulting, custom planting

The Natural Garden
38W443 Highway 64
St. Charles, IL 60175
Ph: (630) 584-0150
Fax: (630) 584-0185
Services: Plants, seeds in limited quantities, consulting, landscape design

INDIANA

J.F. New & Associates
P.O. Box 243
Walkerton, IN 46574
Ph: (219) 586-4300

Spence Nursery
P.O. Box 546
Muncie, IN 47308
Ph: (765) 286-7154

Heartland Restoration Services, Inc.
349 Airport North Office Park
Fort Wayne, IN 46825
Ph: (219) 489-8511

IOWA

Ion Exchange
1878 Old Mission Drive
Harpers Ferry, IA 52146-7533
(319) 535-7231
1-800-291-2143
http://www.ionxchange.com
hbright@polaristel.net
Howard & Donna Bright
(sells plants as well as seeds)

Iowa Prairie Seed Company
1740 220th Street
Sheffield, IA 50475-8031
(515) 892-4111
Daryl Kothenbeutel (sells plants as well as seeds)

Osenbaugh Grass Seeds
RR1 Box 44
Lucas, IA 50151
(515) 766-6476
farm 515/766-6792
home 1-800-LUCAS-88
515/632-8308
John Osenbaugh
Prairie Grass Unlimited, Inc.
P.O. Box 59
Burlington, Iowa 52601
(319) 754-8839

Wildflowers From Nature's Way
RR1 Box 62
Woodburn, IA 50275
(515) 342-6246
Dorothy Baringer

Carl Kurtz
1562 Binford
St. Anthony, IA 50239
(515) 477-8364
(sells seed mixes harvested off native prairie)

McGinnis Tree and Seed Co.
309 E. Florence
Glenwood, IA 51534
(712) 527-4308
(712) 527-4786 fax

Gene Kromray
546 Crestview
Ottumwa, IA 52501
(sells seed mixes harvested off native prairie)

Rose Hill Nursery
2282 Teller Road
Rose Hill, IA 52586
(515) 632-8308

Allendan Seed
1966 175th Ln.
Winterset, IA 50273-8500
Ph: (515) 462-1241
Fax: (515) 462-4084
Services: Seeds, plants, consulting, custom planting, landscaping.

MINNESOTA

Prairie Moon Nursery
Route 3, Box 163
Winona, MN 55987
Ph: (507) 452-1362
Fax: (507) 454-5238
Services: Bare root plants, seeds, publications, consulting, customized seed mixes

Applied Ecology: Native Landscape Restoration and Management.
4316-45th Avenue South,
Minneapolis, MN 55406.
(612) 724-8916,

Feder's Prairie Seed Co.
12871 380th Ave.
Blue Earth, MN 56013-9608.
(507) 526-3049.
feder@blueearth.polaristel.net

"The Prairie Is My Garden" Seed Company,
13633 Ferman Avenue NW,
Clearwater, MN 55320
(612) 878-1694.

Wildlife Habitat.
RR 3, Box 178
Owatonna, MN 55060.
(507) 451-6771

MARYLAND

Environmental Concern Inc,
PO Box P
210 West Chew Avenue
St. Michaels, MD 21663

Numerous flowering plants, shrubs and trees. Some sedges, rushes, grasses.

MASSACHUSETTS

New England Wetland Plants, Inc.
800 Main Street
Amherst, MA 01002
(413) 256-1752
Fax: (413) 256-1092

Red maple, green ash, swamp white oak, black willow, buttonbush, silky dogwood, red-osier dogwood, northern arrowwood, spicebush, winterberry, royal fern, marsh fern, several sedges and rushes and a wetland seed mix among others.

NEW HAMPSHIRE

Gold Star Sod Farm & Nursery, Inc.
250 West Road
Canterbury, NH 03224
(603) 783-4716

Native trees and shrubs including red maple, white birch, swamp azalea, winterberry, sweet pepperbush and others.

NEW YORK

Southern Tier Consulting Inc.
2677 Route 305, P.O. Box 30
West Clarksville, NY 14786
(716) 968-3120 (inquiries)
(800) 848-7614 (orders)

Numerous sedges, rushes, grasses, ferns, shrubs and trees.

VERMONT

New England Nursery Sales
Wally Thrall
PO Box 64
McIndoe Falls, VT 05050
(802) 633-2232

Nursery stockbroker. Sells stock primarily from large nurseries throughout New England and New York.

Northern Nurseries, Inc.
Wholesale Distribution Center
P.O. Box 1048
White River Junction, VT 05001
(802) 295-2117

Distributes plants primarily grown in Connecticut. Wholesale. Trees and shrubs including red maple, sugar maple, white pine, red oak, hemlock, witch hazel, witherod, nannyberry, highbush and lowbush blueberry.

Acorn Ridge Nursery Wholesale Growers

59 Pearl Street
Grand Isle, VT 05458
(820) 372-3811

Swamp azalea, bog rosemary, winterberry, witch hazel, Pagoda dogwood, sheep laurel, sweet fern, bearberry, and wintergreen.

Addison Gardens Wholesale Perennials

RR4, Box 1865
Vergennes, VT 05491
(820) 759-2529

Ginger, marsh marigold, maidenhair fern, lady fern, evergreen wood fern, cinnamon fern, royal fern, Christmas fern.

Cady's Falls Nursery Don & Lela Avery

RD #3, Box 2100
Morrisville, VT 05661
(820) 888-5559

Blue flag iris, water arum, pickerelweed, white water lily, arrowhead, pitcher plants, Braun's holly fern, evergreen wood fern, maidenhair fern, rusty woodsia, bunchberry, American bittersweet, partridgeberry, harebell, wild columbine, red trillium, white trillium.

Carver Family Tree Farm

HCR 32, Box 570 (County Road)
Montpelier, VT 05602
(820) 223-0160

Balsam fir, white spruce, white pine.

Cobble Creek Nursery Wholesale Growers

RD #2, Box 3850
Bristol, VT 05443
(820) 453-3889

Paper birch, red oak, American sycamore, American hornbeam, arrowwood, nannyberry, witherod, highbush cranberry, gray dogwood, red-osier dogwood, witch hazel, dwarf bush honeysuckle, roseshell azalea, elderberry, hobblebush, spiraea, steeplebush, mountain holly, Virginia rose, buttonbush, sweetfern, Virginia creeper, American bittersweet, bearberry.

Drinkwaters Nursery/Landscaping Att: Gary L. Drinkwater

2018 Goss Hollow Road
St. Johnsbury, VT 05819
(820) 748-3317 early a.m. or eve.

Red maple, white birch, gray birch, larch, hemlock, white pine, white spruce, cedar, balsam fir, sugar maple and shadbush.

Elmore Roots Nursery

David L. Fried
P.O. Box 171
Elmore, VT 05657
(820) 888-3305 or 1-800-42PLANT

American beech, white ash, red maple, sugar maple, larch, white cedar, shadbush, highbush cranberry, nannyberry, black chokeberry, Virginia creeper, partridgeberry, and bunchberry.

Evergreen Gardens of Vermont

Carol & David Loysen
P.O. Box 60, Route 100
Waterbury Center, VT 05677
(820) 244-8523

Balsam fir, white pine, hemlock, white spruce, sugar maple, white birch, red maple, red-osier dogwood, winterberry, bog rosemary, elderberry, highbush cranberry, witherod, arrowwood, bearberry, American bittersweet.

Grand Isle Nursery, Inc.

Stephen Spier
P.O. Box 350, 50 Ferry Road
South Hero, VT 05486
(820) 372-8805

Red maple, American beech, butternut, hop hornbeam, larch, Pagoda dogwood, pussy willow, buttonbush, cardinal flower, ostrich fern, and cinnamon fern.

High Reach Farm

Steve Parker & Susanne Terry
2847 Tampico Road
Danville, VT 05828
(820) 748-3512

Sugar maple, red maple, white birch, white ash, yellow birch, black cherry, hop hornbeam, balsam poplar, quaking aspen, balsam fir, white spruce, white pine, hemlock, white cedar, larch, beech, pin cherry, shad wild plum, hobblebush, staghorn sumac, native willows, pagoda dogwood, red-osier dogwood, red elderberry, chokecherry, striped maple, alder, yew, mountain maple, spiraea.

Mettowee Mill Nursery.

Steven Jones
P.O. Box 274, Route 30
Dorset, VT 05251
(820) 325-3007

Red maple, sugar maple, red oak, hemlock, white pine, white cedar, balsam fir, white spruce, white birch, chokeberry; witch hazel, shad, pagoda dogwood, highbush cranberry, winterberry, American bittersweet, Christmas fern maidenhair fern.

Primavera

Tom deGiacomo & Jill Anderson
Box 7
Barnard, VT 05031
(820) 234-5585 Evenings

Butternut, black walnut, sugar maple, swamp white oak, bur oak, chestnut oak, black (sweet) birch, shad, winterberry, hobblebush, swamp azalea, red-stem dogwood, sweet flag, broadleaf arrowhead, softstem bulrush, blue flag ids, pickerelweed, tussock sedge, pitcher plant and foamflower.

Rocky Dale Gardens

62 Rocky Dale Road
Bristol, VT 05443
(820) 453-2782

American beech, red maple, American hornbeam, Pagoda dogwood, witherod viburnum, witch hazel, winterberry, sheep laurel, swamp pink, sweet fern, bush honeysuckle, thimbleberry, lowbush blueberry, cardinal flower, turtlehead, pickerelweed, blue flag iris, columbine, bunchberry, wintergreen, wild ginger, royal fern, cinnamon fern, Christmas fern, lady fern, ostrich fern, maidenhair fern.

South Forty LTD

31 South Forty Road
Shelburne, VT 05482
(820) 985-3351

Sugar maple, silver maple, paper birch, balsam fir, arrowwood viburnum.

U.S. Natural Resources Conservation Service Conservation Districts:

Franklin Cty NRCD (820) 527-1296
Lamoille Cty NRCD (820) 888-4935
Windham Cty NRCD (820) 254-9766
Winooski NRCD (820) 828-4493

Native trees and shrubs are sold at the districts' spring plant sales. Call for a listing and order form.

Vermont Conservation Nursery

Craig Dusablon
P.O. Box 356
Highgate, VT 05459
(820) 868-3861

A conservation packet of 50 plants (6 - 12 inch seedlings): 10 each of white cedar, winterberry, silky dogwood, highbush cranberry, and American bittersweet

WISCONSIN

Wisconsin Natural Re-creation!,

W20531 Gilbo Lane,
Galesville, WI 54630,
(608) 582-2675

Services: Plants, seeds, consulting, custom planting and maintenance

Prairie Nursery

P.O. Box 306
Westfield, WI 53964
Ph: (608) 296-3679
Fax: (608) 296-2741

Prairie Ridge Nursery

RR2, 9738 Overland Road
Mt. Horeb, WI 53572-2832
Ph: (608) 437-5245
Fax: (608) 437-8982
Services: Plants, seeds, consulting, custom planting and maintenance

Taylor Creek Restoration Nurseries

17921 Smith Road
P.O. Box 256
Brodhead, WI 53520-0256
Ph: (608) 897-8641
Fax: (608) 897-8486
E-Mail: Appliedeco@Brodnet.com
Services: Seeds, plants, consulting, custom planting

Country Wetlands Nursery Ltd.

575 W20755 Field Drive
Muskego, WI 53150
(414) 679-1268

Primarily seeds and plants of different sedges, rushes, grasses and flowering plants. Also some shrubs.

Kesters Wild Game Food Nurseries

PO Box 516
Omro, WI 54963
(920) 685-2929
1-800-558-8815 (orders only)

Cattails, burreeds, sedges, sweet flag and other herbaceous plants.

Marshland Transplant Aquatic Nursery

P.O. Box 1
Berlin, WI 54923
(920) 361-4200

A good variety of wet meadow, shallow marsh, deep marsh herbaceous plants. Some trees and shrubs. Also have woodland plants.

Wildlife Nurseries

PO Box 2724
Oshkosh, WI 54903
(920) 231-3780

Sedges, rushes, grasses, cattails, flowering aquatic species.

Appendix C:

A SAMPLER OF GOVERNMENT PROGRAMS

Federal Programs

Most of these programs provide assistance to landowners who own at least 10 acres of land. The Natural Resources Conservation Service and various state agencies supply technical assistance. More information on NRCS programs is available from your local USDA Service Center or on the NRCS web site at www.nrcs.usda.gov/. The amount of money and the type of land that is covered varies from location to location. Some of these programs may not be available in your area.

SIP - The Stewardship Incentive Program. Provides cost-sharing for owners of woodlands. Pays for management plans, tree and shrub planting, forest improvement, windbreaks and hedgerows.

FIP - Forestry Incentives Program - Cost-sharing for private landowners in counties with major industrial forests. Pays for such projects as tree planting, timber stand improvement, preparing sites for natural regeneration.

WHIP - The Wildlife Habitat Improvement Program. Provides cost-sharing to owners of private land not being used for farming. Generally, five acres is the minimum size. Covers projects such as wildlife plantings, prescribed burning, improvement of wildlife and fish habitat. The condition of wildlife habitat on the property adjacent to a landowner's property is also considered when deciding whether to fund a project.

EQIP - The Environmental Quality Incentives Program. Limited to farms. Provides cost-sharing for such projects as prescribed burns, tree and shrub establishment and pruning, wetland development and restoration, wildlife upland habitat management, wildlife watering facilities, habitat management and windbreak establishment and renovation.

CRP - Conservation Reserve Program. Covers highly erodible farmland that was cultivated for at least two of the five years before an application was filed. Rental is paid for the life of the contract; land is taken out of production for at least ten years and planted in grassy or woody cover. . Cost-sharing for wildlife plantings, various erosion control and water-protection projects.

U.S. FISH AND WILDLIFE SERVICE - Partners for Fish and Wildlife Program. Restoration and protection of wildlife and fish habitat on private lands.

CONNECTICUT

Connecticut Forest Stewardship Program
Thomas Worthley, Stewardship Program Forester,
Haddam Cooperative Extension Center,
1066 Saybrook Road,
Box 70,
Haddam CT 06438-0070,
Toll-free phone:
1-888-30WOODS
(1-888-309-6637)
http://www.lib.uconn.edu/CANR/ces/forest/steward.htm

Nature Resources professionals develop a Forest Stewardship plan. Cost-sharing for up to 75% of the cost of plan preparation. You may also be eligible for more money to help carry out the practices recommended on the plan. Requires a minimum of 10 contiguous acres with 75% forest. Cost-sharing might pay for part of forest improvement, wildlife habitat enhancement and other activities.

ILLINOIS

Illinois Department of Natural Resources
524 S. 2nd St., Rm. 400 LTP
Springfield, IL 62701-1787

Maintains two tree nurseries. Landowners who have management plans approved by a District Forester may be eligible to receive seedlings at no cost. See District Forester for information about the Illinois Forestry Development Act.

Acres for Wildlife – The Illinois DNR's Acres for Wildlife program has different requirements for urban and rural participants. Rural landowners must develop at least one acre of land for wildlife, while urban landowners are eligible if land for wildlife totals at least one quarter acre. Participants receive a consultation with a District Wildlife Habitat Biologist, free tree and shrub seedlings and food patch mix, and help in getting financial assistance if it is available for some wildlife management projects.

INDIANA

Indiana Department of Natural Resources
Division of Fish and Wildlife
402 W. Washington St., Rm. W255B
Indianapolis, IN 46204-2748

Game Bird Habitat Program – For owners of at least 10 acres of land. To create or maintain 5-40 acres for game bird food and cover. Cost-sharing up to $100 per acre.

Wildlife Habitat Cost-Share Project – For owners of at least 10 acres of land to develop wildlife habitat. Cost-sharing up to 90 percent of cost of development or $1,000, whichever is less. Help with problem geese. More than 100 birds triggers on-site investigation by district wildlife biologist. Occasionally DNR personnel trap and transport geese during molting. District biologists will help landowners with less than 100 geese obtain a federal permit to trap and move birds and provide information on how to build traps.

Nuisance Wildlife Hotline website, www.anr.ces.purdue.edu/wild.html

IOWA

Department of Natural Resources – Forestry Division
9th and Grand Ave,
Wallace Bldg.
Des Moines, IA 50319-0034

Twelve district foresters help rural and urban landowners with woodland management, insect and disease problems and management of woodland wildlife. See district foresters for management plans and information on cost-sharing.

MAINE

Department of Conservation– Forest Policy and Management Division
22 State House Station
Augusta, ME 04333-0022

Field foresters give technical assistance and information to forest landowners and the general public.

MARYLAND

Department of Natural Resources
Tawes State Office Bldg.
Annapolis, MD 21401
Order trees on the Web
http://www.dnr.state.md.us/forests/nursery/treeinfo.html

Maryland's Wild Acres Program. Gives tips on attracting wildlife to Maryland back yards. Awards certificates to backyard wildlife gardeners. See:

http://www.dnr.state.md.us/wildlife/wildacres.html

MICHIGAN

Department of Natural Resources
Box 30028
Lansing, MI 48909

Nongame wildlife grants between $200 and $5000 are given to private landowners willing to open their land to the general public. The grants are meant to pay for the management or restoration of native plant and animal species. Priority is given to projects where matching funds or volunteer labor and equipment are available.

MINNESOTA

Minnesota Board of Water and Soil Resources plus Soil and Water Conservation Districts
One W. Water St., Suite 200
St. Paul, MN 55107

ReInvest in Minnesota Reserve Program – Protects and restores environmentally sensitive areas such as areas along streams, wetlands and marginal farmlands.

Minnesota Department of Natural Resources – Wildlife Section
500 Lafayette Rd.
St. Paul, MN 55155-4001

Pheasant Habitat Improvement Program – Technical assistance and cost-sharing to landowners for 10-16 row shelterbelts, establishing foodplots, maintaining grassland nesting cover.

NEW JERSEY

The Forest Resource Education Center
NJ Forest Tree Nursery
370 E. Veterans Hwy
Rt. 527-528
Jackson, NJ 08527

The center sells a New Jersey Arbor Day seedling packet consisting of 50 4-inch to 12-inch seedlings of white pine, Norway spruce and a variety of oaks. Landowners who own at least 3 acres of land in New Jersey and who want to plant trees for reforestation and conservation can buy packets of seedlings in multiples of 100.

NEW YORK

Department of Environmental Conservation – Bureau of Private Land Services
50 Wolf Rd.
Albany, NY 12233

Saratoga Tree Nursery – provides trees for conservation planting

Private Forestry Assistance– administers federal cost-share programs, provides information, including ice storm information, for private woodland owners.

OHIO

Department of Natural Resources
Fountain Square
Columbus, OH 43224

Wetland restoration – 50 percent of cost of projects like the construction of small, low-level dikes up to $500 for each acre of restored wetland for a 10-year maintenance agreement or 100 percent of costs up to $1,000 per acre for a 20-year agreement. Contact: 614-265-6907.

VERMONT

Vermont Fish & Wildlife Dept.
103 South Main St.,
Waterbury, VT 05671-0501

The Vermont Nongame and Natural Heritage Program – provides information about Vermont's endangered species, nongame animals, plants and environmentally sensitive areas, and suggestions for land management. Under a separate program, a state wildlife biologist can evaluate suspected deer yards.

WISCONSIN

Wisconsin Department of Natural Resources – Bureau of Forestry
See county forester for details

Wisconsin Forest Landowner Grant Program – Up to 65 percent cost share up to $10,000 a year for management plans and cost of implementation. For at least 10 contiguous acres of non-industrial forest. Includes such projects as tree planting, timber stand improvement, fencing, improvement of wildlife habitat.

Managed Forest Law. – At least 10 acres of wooded land. Forestry plan prepared free by county forester. At least 80 percent must be productive forest land. Reduces property tax on land.

Three forest tree nurseries provide seedlings and transplants.

Other DNR services – Wildlife manager's services for improving or maintaining wildlife habitat.

Appendix D:

SOME PRIVATE GROUPS WITH INFORMATION, EDUCATION AND HELP

BATS

North American Bat Working Groups
http://www.batworkinggroups.org/

Bat Conservation International
P.O. Box 162603
Austin, TX 78716
http://www.batcon.org/

BIRDS

Bluebird Restoration Association of Wisconsin, Inc.
Rt. 1, Box 137-Akron Ave.
Plainfield, WI 54966

Bluebirds Across Vermont Project
255 Sherman Hollow Rd.
Huntinton, VT 05462

Cornell Lab of Ornithology
159 Sapsucker Woods Rd
Ithaca, NY 14850

The Hummingbird Society
P.O. Box 394
Newark, DE 19715

National Bird-Feeding Society
P.O. Box 23
Northbrook, IL 60065-0023

North American Bluebird Society
P.O. Box 74
Darlington, WI 53530-0074

Purple Martin Conservation Association
Edinboro University of Pennsylvania
Edinboro, PA 16444
http://www.purplemartin.org/

BUTTERFLIES, BUGS

Xerces Society
4828 SE Hawthorne Blvd.
Portland, OR 97215

Young Entomologists' Society, Inc.
1915 Peggy Pl.
Lansing, MI 48910-2553
Educational club for teens

GAME BIRDS

National Wild Turkey Federation
770 Augusta Rd.,
P.O. Box 530
Edgefield, SC 29824-0530

Pheasants Forever, Inc.
P.O. Box 75473
St. Paul, MN 55175

Quail Unlimited, Inc.
31 Quail Run
P.O. Box 610
Edgefield, SC 29824-0610

The Ruffed Grouse Society
451 McCormick Rd.
Coraopolis, PA 15108
http://www.ruffedgrousesociety.org/

Wings Over Wisconsin
P.O. Box 202
Mayville, WI 53050

GARDENING INFORMATION

Botanical Club of Wisconsin
c/o Wisconsin Academy of Science, Arts and Letters
1922 University Ave.
Madison, WI 53705

Botanical Society of Western Pennsylvania
5837 Nicholson St.
Pittsburgh, PA 15217-2309

Holly Society of America, Inc.
11318 W. Murdock
Wichita, KS 67212-6609

National Gardening Association
180 Flynn Ave.
Burlington, VT 05401

National Wildflower Research
Center Clearinghouse
4801 La Crosse Ave.
Austin, TX 78739

GENERAL INFORMATION

Environmental Studies at Airlie
7078 Airlie Rd.
Warrenton, VA 20187

Loudoun Wildlife Conservancy
P.O. Box 2088
Purcellville, VA 20132-2088
www.loudounwildlife.org

Merck Forest and Farmland Center, Inc.
P.O. Box 86 - Route 315 Rupert
Mountain Road
Rupert VT 05768
Informational programs at the center

Rachel Carson Council, Inc.
8940 Jones Rd.
Chevy Chase, MD 20815
http://members.aol.com/rccouncil/
ourpage/
Provides information on pesticides

The Wildfowl Trust of North
America, Inc.
P.O. Box 519 Discovery Lane
Grasonville, MD 21638
Education programs at Horsehead
Wetlands Center

GENERAL INTEREST CONSERVATION GROUPS

National Audubon Society
700 Broadway
New York, NY 10003-9501
http://www.audubon.org/
Also many state and local chapters

National Wildlife Federation
8925 Leesburg Pike
Vienna, VA 22184-0001
http://www.nwf.org
Also many state chapters

LANDOWNERS WITH GAME ANIMALS ON THEIR PROPERTY

Farmers and Hunters Feeding
The Hungry
216 North Cleveland Avenue
Hagerstown, MD 21740
www.fhfh.org
Statewide program-deer brought to participating butchers; cash donations cover cost of butchering and packaging

Hunters For The Hungry
P.O. Box 304
Big Island, VA 24526
http://www.h4hungry.org/

Suburban Whitetail Management
of Northern Virginia
P.O. Box 1625
Dale City, VA 22193
http://www.host.trueserver.com/
SWMNV/MainPage.htm
New in 1999. Matches qualified, state licensed bowhunters to non-hunting landowners who are experiencing deer damage in Virginia's Fairfax, Loudoun, or Prince William County. Minimal funding needed comes from organization's founders and from bowhunters' fees.

NATIVE PLANT SOCIETIES

Illinois Native Plant Society
Forest Glen Preserve
20301 E. 900 N. Rd
Westville, IL 61883

Indiana Native Plant and Wildflower
Society
6106 Kingsley Dr.
Indanapolis, IN 46220

Iowa Native Plant Society
Botany Dept.
Iowa State University
Ames, IA 50011-1020

Maryland Native Plant Society
P.O. Box 4877
Silver Spring, MD 20914

Minnesota Native Plant Society
1445 Gortner Ave.
University of Minnesota
St. Paul, MN 55108

Native Plant Society of
Northeastern Ohio
2651 Kerwick
University Heights, OH 44118

Rhode Island Wild Plant Society
12 Sanderson Rd.
Smithfield, RI 02917-2606

Virginia Native Plant Society
P.O. Box 844
Annandale, VA 22003
www.vnps.org

Piedmont Chapter
P.O. Box 336,
The Plains, VA 20198
Publishes "Do I Have to Mow All That? Homeowner's Guide to Preserving our Natural Piedmont Landscape" and "Hedgerow: and Other Comers of Natural Diversity in Our Countryside and Gardens"

Wild Ones - Natural Landscapers, Ltd.
P.O. Box 23576
Milwaukee, WI 53223-0576
http://www.for-wild.org/

Wisconsin Prairie Enthusiasts
4192 Sleepy Hollow Trail
Boscobel, WI 54805

TECHNICAL ASSISTANCE

Chesapeake Wildlife Heritage
P.O. Box 1745
Easton, MD 21601

TRUSTS

Land Trust of Virginia
P.O. Box 354
Leesburg, VA 20178
www.landtrustva.org

Maine Coast Heritage Trust
169 Park Row
Brunswick, ME 04011

The Natural Lands Trust, Inc.
Hildacy Farms
1031 Palmers Mill Rd.
Media, PA 19063
Accepts conservation easements in the Philadelphia metropolitan area

Vermont Land Trust
8 Bailey Avenue
Montpelier, VT 05602
http-//www.vlt.org/
Technical and legal assistance

WOODLANDS

Maryland Forests Assoc.
P.O. Box 599
Grantsville, MD 21536

National Woodland Owners Assoc.
374 Maple Ave., E., Suite 210
Vienna, VA 22180
Also many state chapters

Trees for Tomorrow, Inc. Natural
Resources Education Center
P.O. Box 609
Eagle River, WI 54521
Statewide organization - workshops, sells tree seedlings

Appendix E:

THE TIP OF THE ICEBERG... HOTLISTS AND WEBSITES

BIRDS

Map with some migratory bird sites, plus discussion group possibilities
http://www.gorp.com/gorp/activity/birding/fall.htm

Hotlist of birding web sites
http://www.birder.com/birding/index.html

Listing of relatively current bird sightings in Wisconsin
http://wso.uwgb.edu/wso.htm

Bulletin-board-like discussions on birds, butterflies and wildlife gardens.
http://www.nature.net/forums/

BUTTERFLIES

Hotlist of butterfly web sites
http://www.infocostarica.com/butterfly/news.htm

The Access Indiana Teaching and Learning Center's hotlist of butterfly web sites
http://tlc.ai.org/butterfl.htm

Keep track of migration of monarch butterflies
http://www.MonarchWatch.org/

GOVERNMENT AGENCIES - WEBSITES BY THEM OR ABOUT THEM

hotlist of all cooperative extension services by state
http://www.urbanext.uiuc.edu/Netlinks/ces.html

Hotlist of state natural resources departments
http://www.dfg.ca.gov/links_st.html

U.S. Fish & Wildlife home page
http://www.fws.gov/

U.S. Department of Agriculture list of helpful agencies
http://www.usda.gov/agencies/agencies.htm

Maps of Plant Hardiness Zones
http://www.ars-grin.gov/ars/Beltsville/na/hardzone/ushzmap.html

Links to government and private organizations concerned with plant and animal biodiversity
http://www.heritage.tnc.org/nhp/what.html

Bird identification checklists state by state
http://www.npwrc.usgs.gov/resource/othrdata/chekbird/chekbird.htm

Natural Resources Conservation Service sponsored web site with links to conservation education programs
http://www.fb-net.org/Backyard.htm

University of Wisconsin - Green Bay project surveying bird breeding sites in Wisconsin
http://wso.uwgb.edu/wbba.htm

Information on fur-bearing animals in Illinois
http://www.inhs.uiuc.edu/dnr/fur/

GARDENING

*Includes excellent illustrated directory with factual information about seeds for sale
http://www.prairiefrontier.com/index.html

Time-Life Plant Encyclopedia, with link to garden.com where plants can be purchased
http://www.vg.com/cgi-bin/v2/gema/PID=69909268174122463813336, 13336&s=4393

Hotlist of gardening web sites, including sites where plants and seeds can be purchased
http://www.webring.org/cgi-bin/webring9ring=geogarden&list

Listing of publicly and privately owned gardens open to the public throughout the U.S. and Canada
http://www.botanique.com/tourmast.htrnl

SOURCES FOR REPELLING WILDLIFE

*Lifestyle information about problem birds, plus products to repel them
http://www.birdbarrier.com/home.html

*products only to repel problem birds
http://www.thomasregister.com/olc/birdbgone/

*New Jersey service brings in trained border collies to repel unwanted Canada geese
http://www.njgeese.com/

USDA- APHIS (Animal and Plant Health Inspection Service)
http://www.aphis.usda.gov/ws/pubs.html

*protective sleeves for young trees, flagging to discourage geese, animal traps, repellents and repelling devices
http://www.hartpub.com/bg/v3/8e.htm
Also available at
Ben Meadows Co.
3589 Broad St.
Atlanta, GA 30341
(405) 241-6401

*protective sleeves for young trees, planting supplies
http://www.usmmag.com/free_advertisers/forestry_suppliers.htm
Also available at
Forestry Suppliers
205 W. Rankin St.
Jackson MS 39201
Tel: (800)647-5368

*commercial sites with information about products to buy. Unmarked sites may also be commercial sites, but information is not primarily about items for sale.

BIBLIOGRAPHY

Sources

1. AAA Company, AAA website, http://www.aaawildlife.com, 1999.
2. Armstrong, Pat, *Wild Ones Handbook: Prairie Plants Evolved to a Harsh Climate,* Wild Ones, undated.
3. Bailey, Stephanie, *How to Make Butterfly Gardens,* University of Kentucky College of Agriculture website, http://www.uky.edu/Agriculture/Entomology/entfacts/misc/ef006.htm , undated.
4. Benyus, Janine M., *The Field Guide to Wildlife Habitats of the Eastern United States,* Fireside (Simon and Schuster), New York, NY, 1989.
5. Bittner, Steve, *Western Maryland Black Bears,* Maryland Department of Natural Resources, Wildlife and Heritage Division, updated May 7, 1999.
6. Blodgett, Doug, "Wild Turkey" in *A Landowner's Guide, Wildlife Habitat Management for Vermont Woodlands,* Vermont Agency of Natural Resources, 1995.
7. Bluett, Robert, *Nuisance Raccoons in Urban Settings,* Illinois Dept. of Conservation, Division of Wildlife Resources, September, 1992.
8. Bluett, Robert and Scott Craven, *The Raccoon,* Bureau of Wildlife Management, Wis. Dept. of Natural Resources, January 1985
9. Bogenschutz, Todd, *Pheasant Food Plots — What exactly do they do?,* Pheasants Forever magazine, Winter, 1993.
10. Booth, Mary and Melody Mackey Allen, "Butterfly Garden Design" in *Butterfly Gardening: Creating Summer Magic in Your Garden,* Sierra Club Books, San Francisco, 1998.
11. Brewer, Jo, "Notes from a Butterfly Gardener" in *Butterfly Gardening: Creating Summer Magic in Your Garden,* Sierra Club Books, San Francisco, 1998.
12. Brownlee, David H., Northern New England Animal Damage Control Program Education Leaflet Series--Porcupine, University of Massachusetts, Amherst, MA, undated.
13. Buck, John, "Woodcock" in *A Landowner's Guide, Wildlife Habitat Management for Vermont Woodlands,* Vermont Agency of Natural Resources, 1995.
14. Bureau of Endangered Resources, Wisconsin Dept. of Natural Resources, Wisconsin Manual of Control Recommendations for Ecologically Invasive Plants, Bureau of Endangered Resources, Wisconsin Dept. of Natural Resources, May, 1997.
15. Buskirk, Daniel D., Glenn R. Dudderar and Harland D. Ritchie, *Deer Barriers . . . fencing, repellents, dog restraint systems, scare devices,* Michigan State University Extension Bulletin E-2672, August, 1998.
16. Cabela, Cabela's Master Catalog, Fall 1999, Edition 1, , 1999.
17. Cadieux, Charles L., *Wildlife Management on Your Land, the Practical Owner's Manual on How, What, When and Why,* Stackpole Books, Harrisburg PA, 1985.
18. Chadwick, Nan, *Controlling Wildlife Damage — Skunks,* University of Massachusetts Cooperative Extension Service, November, 1984.
19. Chadwick, Nan, *Controlling Wildlife Damage — Woodpeckers,* University of Massachusetts Cooperative Extension Service, November, 1984.
20. Chadwick, Nan, *Controlling Wildlife Damage — Woodchuck,* University of Massachusetts Cooperative Extension Service, November, 1984.
21. Chadwick, Nan, *Wildlife in Massachusetts — Raccoon,* University of Massachusetts Cooperative Extension Service, 1984.
22. Chambers, Lanny, *Hummingbirds!* website, Lanny Chambers, http://www.hummingbirds.net, 1999.
23. Coverstone, Nancy, with Catherine A. Elliott and Judy Walker, *Habitats: Components of a Backyard Wildlife Habitat* (Bulletin #7137), University of Maine Cooperative Extension, 4/98.
24. Coverstone, Nancy, with Catherine A. Elliott and Judy Walker, *Habitats: Principles for Creating a Bakyard Wildlife Habitat* (Bulletin # 7132), University of Maine Cooperative Extension, 4/98.
25. Craven, Scott, *Protecting Gardens and Landscape Plantings from Rabbits,* University of Wisconsin Extension, 1996.

26. Craven, Scott, *Skunks: How to Deal with Them,* University of Wisconsin Extension, 1998.

27. Crenshaw, Bill, "Gray Squirrel" in *A Landowner's Guide, Wildlife Habitat Management for Vermont Woodlands,* Vermont Agency of Natural Resources, 1995.

28. Decker, Daniel J. and John W. Kelley, *Enhancement of Wildlife Habitat on Private Lands,* Media Services, Cornell University, 7/86, rev. 2/98.

29. DeGraaf, Richard M. and Alex L. Shigo, *Managing Cavity Trees for Wildlife in the Northeast* GTR NE-101, USDA Forest Service, Northeastern Forest Experiment Station, 1985.

30. DeGraaf, Richard M. and David A. Richard, *Forest Wildlife of Massachusetts,* University of Massachusetts Cooperative Extension Service, 1987.

31. Denny, Guy L., *Introduction to Ohio's Trees,* Ohio Department of Natural Resources, undated.

32. DeStefano, Stephen, Scott R. Craven, Robert L. Ruff, and John F. Kubisiak, *A Landowner's Guide to Woodland Wildife Management,* Universityof Wisconsin Cooperative Extension, 1994

33. Dix, Edward T., editor, *Common Trees of Pennsylvania,* Pennsylvania Dept. of Conservation and Natural Resources, undated.

34. Dudderar, Glenn, *Nature from Your Back Door,* Outreach Communications, Michigan State University, 1991.

35. Ehrlich, Paul R., David S. Dobkin, and Darryl Wheye, *The Birder's Handbook,* Simon & Schuster, New York, 1988.

36. Elliott, Catherine A., *Wild Apple Trees for Wildlife,* University of Maine Cooperative Extension, March, 1998.

37. Ellis, Barbara, *Attracting Birds & Butterflies,* Houghton Mifflin Company, Boston & NY, 1997.

38. Fix, W. L, Extension Forester, *Planting Forest Trees and Shrubs in Indiana,* agcom.purdue.edu/AgCom/Pubs/FNR/FNR-36.html, .

39. Garland, Larry, "Technical Assistance for Vermont Landowners" in *A Landowner's Guide, Wildlife Habitat Management for Vermont Woodlands,* Vermont Agency of Natural Resources, 1995.

40. Golden, Ed, *Making a home for Wild Turkeys,* Maryland Department of Natural Resources, Wildlife and Heritage Division.

41. Guiterrez, R. J., D.J. Decker, R. A. Howard, JR., and J.P. Lassoie, *Managing Small Woodlands for Wildlife,* Cornell University, Ithaca, NY, 1984.

42. Gullion, Gordon W., *Managing Northern Forests for Wildlife,* The Ruffed Grouse Society, 1984.

43. Hassinger, Jerry, Lou Hoffman, Michael J. Puglisis, Terry D. Rader and Robert G. Wingard, *Woodlands and Wildlife,* Penn State College of Agricultural Sciences, 1979.

44. Henderson, Carrol L., *Landscaping for Wildlife,* Minnesota Dept. of Natural Resources, 1987.

45. Henderson, Carrol L., *Wild About Birds,* Minnesota's Bookstore, St. Paul, MN, 1995.

46. Hunn, David, "No. 761 and counting Biologists on the hunt for urban bear understanding", *Anchorage Daily News,* August 15, 1999.

47. Hyland, Fay, *The Conifers of Maine,* University of Maine Cooperative Extension Service, undated.

48. Illinois Dept. of Natural Resources, *Landscaping for Wildlife,* Ill. Dept. of Natural Resources - Natural Heritage Division, undated.

49. Illinois Dept. of Natural Resources, Division of Natural Heritage, *Wood Projects for Illinois Wildlife,* Illinois Dept. of Natural Resources, Division of Natural Heritage, undated.

50. Indiana Dept. of Natural Resources, Division of Fish and Wildlife, *Life Series-The Barn Owl,* Indiana Dept. of Natural Resources, Division of Fish and Wildlife, 1996.

51. Indiana Dept. of Natural Resources, Division of Fish and Wildlife, *Life Series-Bobwhite Quail,* Indiana Dept. of Natural Resources, Division of Fish and Wildlife, 1997.

52. Indiana Dept. of Natural Resources, Division of Fish and Wildlife, *Managing Deer Damage,* Indiana Dept. of Natural Resources, Division of Fish and Wildlife, undated.

53. Iowa Department of Natural Resources, *Attracting Backyard Wildlife,* Iowa Department of Natural Resources,

54. Judd, Mary K., Project Director, *Gimme Shelter — Shelterbelts and Food Plots for Wildlife,* Bureau of Wildlife Management, Wis. Dept. of Natural Resources, undated.

55. Judd, Mary K., Project Director, *To Cut or Not to Cut? Managing Your Woodland for Wildlife,* Bureau of Wildlife Management, Wis. Dept. of Natural Resources, undated.

56. Judd, Mary K., Project Director, *Home on the Range — Restoring and Maintaining Grasslands for Wildlife,* Bureau of Wildlife Management, Wis. Dept. of Natural Resources, undated.

57. Judd, Mary Kay, Diane Schwartz, Todd Peterson, *Wildlife and Your Land: On Edge,* Bureau of Wildlife Management, Wis. Dept.of Natural Resources, undated.

58. Judd, Mary Kay, Diane Schwartz, Todd Peterson, *Wildlife and Your Land: Putting Pen to Paper,* Bureau of Wildlife Management, Wis. Dept.of Natural Resources, undated.

59. Kashanski, Catherine, *Native Vegetation for Lakeshores, Streamsides and Wetland Buffers,* Vermont Agency of Natural Resources, September, 1996.

60. Kays, Jonathan, *Controlling Deer Damage in Maryland,* University of Maryland Cooperative Extension Service, Bulletin 354, undated.

61. Koelling, Mel, *Identifying Trees of Michigan,* Michigan State University Extension Service, October, 1997.

62. Krasny, Marianne E., *Wildlife in Today's Landscapes,* Cornell Cooperative Extension, 2/91.

63. Kress, Stephen W., Guest Editor, *Bird Gardens: Welcoming Wild Birds to Your Yard,* Brooklyn Botanic Garden Publications, Autumn, 1998.

64. Krischik, Vera, *Butterfly Gardening,* University of Minnesota Extension Service, St. Paul, MN, 1996.

65. Kubisiak, John, *Oak Forests: A Management Opportunity for Ruffed Grouse and Other Wildlife,* Ruffed Grouse Society, Coraopolis, PA, 1987.

66. Kundt, John F. and Robert L. Baker, *Leaf Key to Common Trees in Maryland,* University of Maryland Cooperative Extension Service, 1983-4.

67. Leedy, Daniel L., and Lowell W. Adams, A *Guide to Urban Wildlife Management,* National Institute for Urban Wildlife, 1984.

68. Leopold, A. Starker, Ralph Guitierrez and Michael T. Bronson, *North American Game Birds and Mammals,* Charles Scribner's Sons, New York, NY, 1981.

69. Lewis, Alcinda, Guest Editor, *Butterfly Gardens: Luring Nature's Loveliest Pollinators to Your Yard,* Brooklyn Botanic Garden Publications, Summer, 1997.

70. Lobner, Eric, Editor, *Managing Your Land for Wild Turkeys,* Wisconsin Dept. of Natural Resources, July, 1998.

71. Lobner, Eric, Editor, *Wisconsin Turkey Hunter's Guide,* Wisconsin Dept. of Natural Resources, Wisconsin Chapter of the National Wild Turkey Federation, December, 1998.

72. Martin, Alexander, Herbert S. Zim and Arnold L. Nelson, *American Wildlife & Plants: A Guide to Wildlife Food Habits,* Dover Publications, 1951.

73. Maryland Department of Natural Resources, Wildlife and Heritage Division, *When Resident Geese Become a Problem,* Maryland Department of Natural Resources, Wildlife and Heritage Division, undated.

74. Maryland Dept. of Natural Resources, *When Resident Geese Become a Problem,* Maryland Dept. of Natural Resources website, 1999.

75. McClain, William E., *Prairie Establishment and Landscaping,* Division of Natural Heritage, Illinois Dept. of Natural Resources, Springfield, Ill., 1997.

76. Minnesota Department of Natural Resources, *Woody Cover Plantings for Wildlife,* Minnesota Department of Natural Resources, 1989.

77. Minnesota Dept. of Natural Resources, Forestry Division, *Trees and Large Shrubs: Species Native to Minesoata's Ecological Regions,* Minnesota Dept. of Natural Resources, Forestry Division, 1995.

78. Montague, Fred H. , Jr., *Wildlife Field Notes: How Vines Provide Wildlife with Food and Cover,* Purdue University (agcom.purdue.edu/AgCom/Pubs/FNR/FNR-96.html) Dept. of Forestry and Natural Resources, .

79. Nelson, Richard, *Heart and Blood: Living with Deer in America,* Vintage Books (Random House), New York, NY, October, 1998.

80. New Jersey Dept. of Environmental Protection & Energy, Div. of Parks & Forestry, Forestry Services, *Trees to Attract Wildlife,* New Jersey Dept. of Environmental Protection & Energy, Div. of Parks & Forestry, Forestry Services, undated.

81. New Jersey Dept. of Environmental Protection & Energy, Div. of Parks & Forestry, Forestry Services, *Deer Damage on Ornamental Plants,* New Jersey Dept. of Environmental Protection & Energy, Div. of Parks & Forestry, Forestry Services, .

82. New York State Dept. of Environmental Conservation & the New York State Chapter of the National Wild Turkey Federation, *The Wild Turkey in New York,* New York State Dept. of Environmental Conservation & the New York State Chapter of the National Wild Turkey Federation, 9/98.

83. Norrgard, Ray and members of the Farmland Wildlife Committee, *Establishing and Managing Nesting Cover for Wildlife,* Minnesota Department of Natural Resources, 1989.

84. Oskay, Greg, *Landscaping for Wildlife — Native Shrubs and Vines for Backyard Wildlife Habitats,* Water Resources Design, Inc. 551 Teton Trail, Indianapolis, IN 46217, 1998.

85. Oskay, Greg, *Landscaping for Indiana Wildlife — Native Trees for Backyard Wildlife Habitats,* Water Resources Design, Inc. 551 Teton Trail, Indianapolis, IN 46217, March, 1998.

86. Parkhurst, James A., *Northern New England Animal Damage Control Program Education Leaflet Series — Moles,* University of Massachusetts, Amherst, MA, undated.

87. Parren, Steve, *Backyard Wildlife Habitat in Vermont,* Vermont Department of Fish and Wildlife, 1997.

88. Parren, Steve, "Habitat Concepts and Features" in *A Landowner's Guide, Wildlife Habitat Management for Vermont Woodlands,* Vermont Agency of Natural Resources, 1995.

89. Payne, Neil and Fred C. Bryant, *Wildlife Habitat Management of Forestlands, Rangelands and Farmlands,* Krieger Publishing Company, 1998 .

90. Pennsylvania State University, College of Agricultural Sciences, *Ornamental Fruited Trees & Shrubs in Penna.,* Pennsylvania State University, College of Agricultural Sciences, undated.

91. Pheasants Forever, *Establishing Food and Cover Plots,* Pheasants Forever, undated.

92. Pheasants Forever, *The Facts of Pheasant Stocking,* Pheasants Forever, undated.

93. Pheasants Forever, *Guidelines for Establishing Warm Season Vegetation,* Pheasants Forever, undated.

94. Prairie Nursery, 1999 catalog, Prairie Nursery, P.O. Box 306, Westfield, WI 53964, 1999.

95. Prince Bird Food Corporation, *How to Attract Birds to Your Backyard,* Prince Corp., 8351 County H, Marshfield, WI 54449, undated.

96. Pyle, Robert Michael, "Butterfly Watching Tips," in *Butterfly Gardening: Creating Summer Magic in Your Garden,* Sierra Club Books, San Francisco, 1998.

97. Reay, Russell S., Douglas W. Blodgett, Barbara S. Burns, Steven J. Weber and Terry Frey, *Management Guide for Deer Wintering Areas in Vermont,* Vermont Dept. of Fish and Wildlife and Dept. of Forests, Parks and Recreation, September, 1990.

98. Regan, Ron, "Physiographic Regions of Vermont" in *A Landowner's Guide, Wildlife Habitat Management for Vermont Woodlands,* Vermont Agency of Natural Resources, 1995.

99. Robinson, Sandra, Black Bear, University of Massachusetts, Amherst, MA, December, 1992.

100. Rothschild, Miriam, "What Do Butterflies See?" *Butterfly Gardening: Creating Summer Magic in Your Garden,* Sierra Club Books, San Francisco, 1998.

101. Rothschild, Miriam, "Gardening with Butterflies" in *Butterfly Gardening: Creating Summer Magic in Your Garden,* Sierra Club Books, San Francisco, 1998.

102. Royar, Kim, "Eastern Cottontail Rabbit" in *A Landowner's Guide, Wildlife Habitat Management for Vermont Woodlands,* Vermont Agency of Natural Resources, 1995.

103. Salwey, Mary K., Janet L. Hutchens, Todd Peterson and others, *Wisconsin Wildlife Primer,* Bureau of Wildlife Management, Wis. Dept. of Natural Resources, 1998.

104. Seely, Ron, "Botanist makes case for thinning state deer herd", *Wisconsin State Journal,* September 18, 1999.

105. Smith, Arthur E., Scott R. Craven and Paul D. Curtis, *Managing Canada Geese in Urban Environments,* Cornell University Cooperative Extension, Ithaca, N.Y., 1999.

106. State of Vermont, Fish and Game Dept., *A Landowner's Guide, Wildlife Habitat Management for Vermont Woodlands,* State of Vermont, Fish and Game Dept., 1979.

107. Stukel, Eileen Dowd, Douglas C. Backlund, Maggie E. Hachmeister and Terry Wright, *Sharing Your Space,* Wildlife Division, South Dakota Dept. of Game, Fish and Parks, 1995.

108. Taylor, Norman, editor, *Taylor's Encyclopedia of Gardening,* 4th edition, Houghton Mifflin Co., Boston, 1976.

109. Townsend, Thomas W., *Control of Deer Damage in Tree Plantations,* Ohio State University Extension, undated.

110. Tubbs, Carl, Richard M. DeGraaf, Mariko Yamasaki, and William M. Healy, *Guide to Wildlife Tree Management in New England Northern Hardwoods,* United States Department of Agriculture, 1986.

111. Tufts, Craig and Peter Loewer, *Gardening for Wildlife,* Rodale Press, Emmaus PA, 1995.

112. University of Maine Cooperative Extension, *Birdhouse Basics,* University of Maine Cooperative Extension, February, 1999.

113. U.S. Fish and Wildlife Service, *Homes for Birds,* Dept. of the Interior, US Fish and Wildlife Service, 1991.

114. U.S. Fish and Wildlife Service, *Backyard Bird Feeding,* Dept. of the Interior, US Fish and Wildlife Service, 1990.

115. U.S. Fish and Wildlife Service, *Attract Birds,* Dept. of the Interior, US Fish and Wildlife Service, 1991.

116. USDA Soil Conservation Service, Conservation Choices. *Your Guide to 30 Conservation and Environmental Farming Practices,* USDA SCS, Madison, WI, 1994.

117. USDA, Natural Resources Conservation Service, *Wildlife Food Plots,* Wisconsin Job Sheet 136, USDA, Natural Resources Conservation Service, undated.

118. Van Hoey, Alger F., *Wildlife Management with Herbaceous Cover,* Management Series No. 4, Indiana Dept. of Natural Resources, Division of Fish and Wildlife, undated.

119. Venable, Norma Jean, *Selected Trees and Shrubs of West Virginia,* West Virginia University Extension Service, 1995.

120. Willey, Charles, "Black Bear" in *A Landowner's Guide, Wildlife Habitat Management for Vermont Woodlands,* Vermont Agency of Natural Resources, 1995.

121. Winter, Dave, "The Struggle to Survive" in *Butterfly Gardening: Creating Summer Magic in Your Garden,* Sierra Club Books, San Francisco, 1998.

122. Wisconsin DNR Bureaus of Forestry and Wildlife Management, *Windbreaks that Work!,* Wisconsin DNR Bureaus of Forestry and Wildlife Management, .

123. Wisconsin DNR Bureaus of Forestry and Wildlife Management, *Woody Cover for Wildlife,* Wisconsin DNR Bureaus of Forestry and Wildlife Management, undated.

124. Wise, Sherry, *The Cottontail Rabbit,* Bureau of Wildlife Management, Wis. Dept. of Natural Resources, February 1986.

125. Wise, Sherry, *The Fox and Gray Squirrels,* Bureau of Wildlife Management, Wis. Dept. of Natural Resources, 1986.

126. Wise, Sherry, *The Gray Partridge,* Bureau of Wildlife Management, Wis. Dept. of Natural Resources, 1986.

127. Wise, Sherry, *The Ring-Necked Pheasant,* Bureau of Wildlife Management, Wis. Dept. of Natural Resources, March, 1986.

128. Wise, Sherry, *The Snowshoe Hare,* Bureau of Wildlife Management, Wis. Dept. of Natural Resources, February 1986.

129. Wise, Sherry, *The Black Bear,* Bureau of Wildlife Management, Wis. Dept. of Natural Resources, March, 1986.

130. Wise, Sherry, *The Bobwhite Quail,* Bureau of Wildlife Management, Wis. Dept. of Natural Resources, 1988.

131. Wise, Sherry and John Kubisiak, *The Ruffed Grouse,* Bureau of Wildlife Management, Wis. Dept. of Natural Resources, February, 1986.

132. Yorke, Diane E., *Wildlife Habitat Improvement: Farmlands and Wildlife,* University of New Hampshire Cooperative Extension, Sept. 1995.

133. Zimmerman, Dean, *Controlling Nuisance Game Mammals,* Management Series No. 10, Indiana Dept. of Natural Resources, Division of Fish and Wildlife, undated.

134. Zimmerman, Dean, *Controlling Nuisance Non-Game Wildlife,* Management Series No. 9, Indiana Dept. of Natural Resources, Division of Fish and Wildlife, undated.

Tufted Titmouse
(Patricia Gilbert)

TABLE 3-1

COMMON BIRDS BY STATE

	CT	DE	IL	IN	IA	ME	MD	MA	MI	MN	NH	NJ	NY	OH	PA	RI	VT	VA	WI
Mourning Dove		*	*	*	*	*	*	*	*	*	*	*	*	*	*	*	*	*	*
Yellow-billed Cuckoo	*	*	S		NE		ES	*	*					*				*	C,W
Barn Owl		*						*							SE				
Eastern Screech Owl	*	*	N				*	*	*				*						S
Great Horned Owl	*	*	*	*	*		*	*	*		*	C	C	N	*		*	*	*
Barred Owl	*	*	C	*	NE	C	C	*	*		*	C			NW			*	*
Common Nighthawk	*	*	C		NE			*	*	*	*	C		C	SE			*	C,W
Whippoorwill		*	C		NE			*	*				LI						C
Chimney Swift	*	*	N,S		NE	C	C	*	*	*	*	C		*	*			*	W
Ruby-throated Hummingbird	*	*	C				*	*	*	*	*				SE			*	
Belted Kingfisher		*	C,N		NE	C	*	*	*	*	*	C	*	N	*	*	*	*	C,W
Red-headed Woodpecker		*	C	*	*			*	*					N				*	C,W
Red-bellied Woodpecker	*	*	C,S	*	*		*	*	*	*		C	LI	C				*	W

	CT	DE	IL	IN	IA	ME	MD	MA	MI	MN	NH	NJ	NY	OH	PA	RI	VT	VA	WI
Yellow-bellied Sapsucker	*		C		NE	C		*	*	*				N	NW		*	*	W
Downy Woodpecker	*	*	N,S	*	*	*	*	*	*	*	*	C	*	*	*		*	*	*
Hairy Woodpecker	*			*	NE	C		*	*	*	*	C					*	*	*
Northern Flicker	*	*	*	*	*	*	*	*	*	*	*	C	*	*	*	*	*	*	*
Pileated Woodpecker			C				*	*	*									*	W
Eastern Wood Peewee	*	*	*	*	NE	C	*	*	*		*	C	LI	*	NW		*	*	*
Least Flycatcher	*		C		NE	C		*	*	*				N	NW		*	*	*
Eastern Phoebe	*	*	C,S	*	NE	C	*	*	*	*	*	C	*	N	NW		*		*
Great-crested Flycatcher		*	C,S		NE	C	*	*			*	C	LI	*	NW		*	*	*
Eastern Kingbird	*	*	*	*	NE	*	ES	*	*		*	C	*	*	*		*	*	*
Horned Lark		*	C		NE			*	*					N				*	S,C
Purple Martin	*	*	*	*	NE		ES	*	*				C	N	NW			*	
Tree Swallow	*	*	*	*	*	*	ES	*	*	*	*	*	*	N	*	*	*	*	*
Barn Swallow	*	*	*	*	*	*	ES	*	*	*	*	*	C	*	*	*	*	*	*
Blue Jay	*	*	*	*	*	C	*	*	*	*	*	*	*	*	*	*	*	*	*
American Crow	*	*	*	*	*	*	*	*	*	*	*	C	*	C	*	*	*	*	*
Black-capped Chickadee	*		C,N		*	*		*	*	*	*	C	*		NW	*	*	*	*
Tufted Titmouse	*	*	C,S	*	NE		*	*	*		*	C	LI	*	SE	*		*	

	CT	DE	IL	IN	IA	ME	MD	MA	MI	MN	NH	NJ	NY	OH	PA	RI	VT	VA	WI
White-breasted Nuthatch	*		C,N	*	*		*	*	*	*	*	C	*	*	NW	*	*	*	*
Brown Creeper	*		C		NE			*	*	*	*	C		N	SE		*	*	W
House Wren	*	*	*		*		*		*	*	*	C	*	*	*	*		*	*
Marsh Wren	*	*			NE		ES	*	*			*	*	N	SE	*		*	S,W
Golden-crowned Kinglet			C,N			*	ES	*	*	*	*		C	N				*	S
Ruby-crowned Kinglet	*		C,N		NE	CO	ES	*	*	*	*	C	C	N	SE			*	*
Eastern Bluebird	*		S		NE		ES	*	*	*		C		C				*	W
Veery	*				NE	C	C	*	*		*	C	C		*		*	*	C,W
Wood Thrush	*	*	C		NE		C	*	*		*	C	*	*	*		*	*	W
American Robin	*	*	*	*	*	*	*	*	*	*	*	*	*	*	*	*	*	*	*
Gray Catbird	*	*	C	*	*	CO	*	*	*	*	*	*	*	*	*	*	*	*	*
Northern Mockingbird	*	*	S	*			*	*	*		*	*	LI		SE	*		*	
Brown Thrasher	*	*	C,S	*	*		ES	*	*			C		*	SE	*		*	C,W
Cedar Waxwing	*	*	*		NE	*		*	*	*	*	C		*	NW	*	*	*	*
European Starling	*	*	*	*	*	*	ES	*	*	*	*	*	*	*	*	*	*	*	S,W
Warbling Vireo	*		C,S	*	NE	C		*	*	*	*		C	*	SE				S,W
Red-eyed Vireo	*		C,N		NE	*	*	*	*	*	*	C	C	*	*		*	*	*
Tennessee Warbler	*		C,N		NE			*	*	*	*	C		N	NW			*	*

	CT	DE	IL	IN	IA	ME	MD	MA	MI	MN	NH	NJ	NY	OH	PA	RI	VT	VA	WI
Nashville Warbler			C,N		NE	*		*	*	*	*		C	N	NW				*
Yellow Warbler	*		C,N	*	*	*	*	*	*	*	*	*	LI	*	*	*	*	*	*
Chestnut-sided Warbler	*		C			C		*	*		*	C			*			*	S
Magnolia Warbler			C,N			*		*	*		*	C	C	N	*			*	S
Yellow-rumped Warbler	*	*	C,N	*	*	*	ES	*	*	*	*	*	*	*	*	*		*	*
Black-and-White Warbler	*	*	C,N		NE	*	C	*	*	*	*	*	*	N	SE			*	S,W
American Redstart	*	*	C,N		*	*	C	*	*	*	*	C	*	N	*		*	*	S,W
Ovenbird	*	*	C			C	*	*	*	*	*	C	*		*			*	C
Northern Water Thrush	*		C			C		*	*		*				SE			*	
Common Yellowthroat Warbler	*	*	*	*	W	*	*	*	*	*	*	*	*	*	*	*	*	*	*
Scarlet Tanager	*	*					*	*	*		*	C			*			*	C
Northern Cardinal	*	*	*	*	*		*	*	*	*	*	*	*	*	*	*		*	S,W
Rose-breasted Grosbeak	*		C		*	C		*	*	*	*	*	C		NW		*		*
Indigo Bunting	*	*	*	*	NE		*	*	*	*	*		C	*	*			*	*
Rufous-sided Towhee		*	C,S	*	W		*		*		*	C	*	C	SE	*		*	C

	CT	DE	IL	IN	IA	ME	MD	MA	MI	MN	NH	NJ	NY	OH	PA	RI	VT	VA	WI
American Tree Sparrow	*		*	*	*	C	C	*	*	*	*	C	*	N	*		*	*	*
Chipping Sparrow	*	*	C,N	*	NE	C	*	*	*	*	*	C	*		NW		*	*	C,W
Field Sparrow	*	*	C,N	*	*		C	*	*	*	*	C	C	*	*			*	C,W
Fox Sparrow	*		C					*	*	*	*		C	N				*	S
Song Sparrow	*	*	C,N	*	*	C	C	*	*	*		*	*	*	*	*	*	*	*
White-throated Sparrow	*	*	*	*	NE	C	*	*	*	*	*	*	*	*	*		*	*	*
House Sparrow	*	*	*	*	*	*	*	*	*	*	*	*	*	*	*		*	*	S
Dark-eyed Junco	*	*	*	*	*	*	*	*	*	*	*	*	LI	*	*	*	*	*	*
Bobolink	*					C		*	*	*	*			N			*	*	S
Red-winged Blackbird	*	*	*	*	*	*	ES	*	*	*	*	*	*	*	*	*	*	*	*
Eastern Meadowlark	*	*	*	*	*		ES	*	*	*	*		C	N	NW	*	*	*	S
Common Grackle	*	*	*	*	*	*	*	*	*	*	*	*	*	*	*	*	*	*	*
Brown-headed Cowbird	*	*	N,S	*	*	C	*	*	*	*	*	*	C	*	*	*	*	*	*
Baltimore Oriole	*				NE			*	*	*	*	C	C	N	*		*	*	C,W
Purple Finch	*		S		NE	C	C	*	*	*	*		C	C					N
House Finch	*	*	N,S				C	*	*	*	*	*	LI	C	SE			*	
American Goldfinch	*	*	N,S	*	*	C		*	*	*	*	*	C	*	*	*	*	*	*

This table is a compilation of checklists that appear on The Northern Prairie Wildlife Research Center's *Bird Checklists of the United States*.

The common names of the birds and the order in which they appear on the list are according to the sixth American Ornithologists' Union *Checklist of North American Birds*. Birds listed as common or abundant in a state during at least two seasons of the year are indicated with a star (*) on the chart. A star also means that the bird has been sighted in a variety of places in that state. Where a bird has been seen often in or near only one wildlife refuge in a state, the section of the state is indicated by a letter. Where a box is empty, the bird is rare to nonexistent in that state, or the bird is only present in significant numbers during one season of the year. In other words, someone might be seeing that bird in your state, but the odds are that you won't.

The letter code:

N=north
LI=Long Island (NY)
ES=Eastern Shore (MD)
S=south
NE=northeast

SE=southeast
NW=northwest
W=west
CO=coastal
C=central or non-coastal

White Breasted Nuthatch
(Patricia Gilbert)

TABLE 3-2

BIRDS' HABITATS AND NATURAL FOODS

Chickadee (Patricia Gilbert)

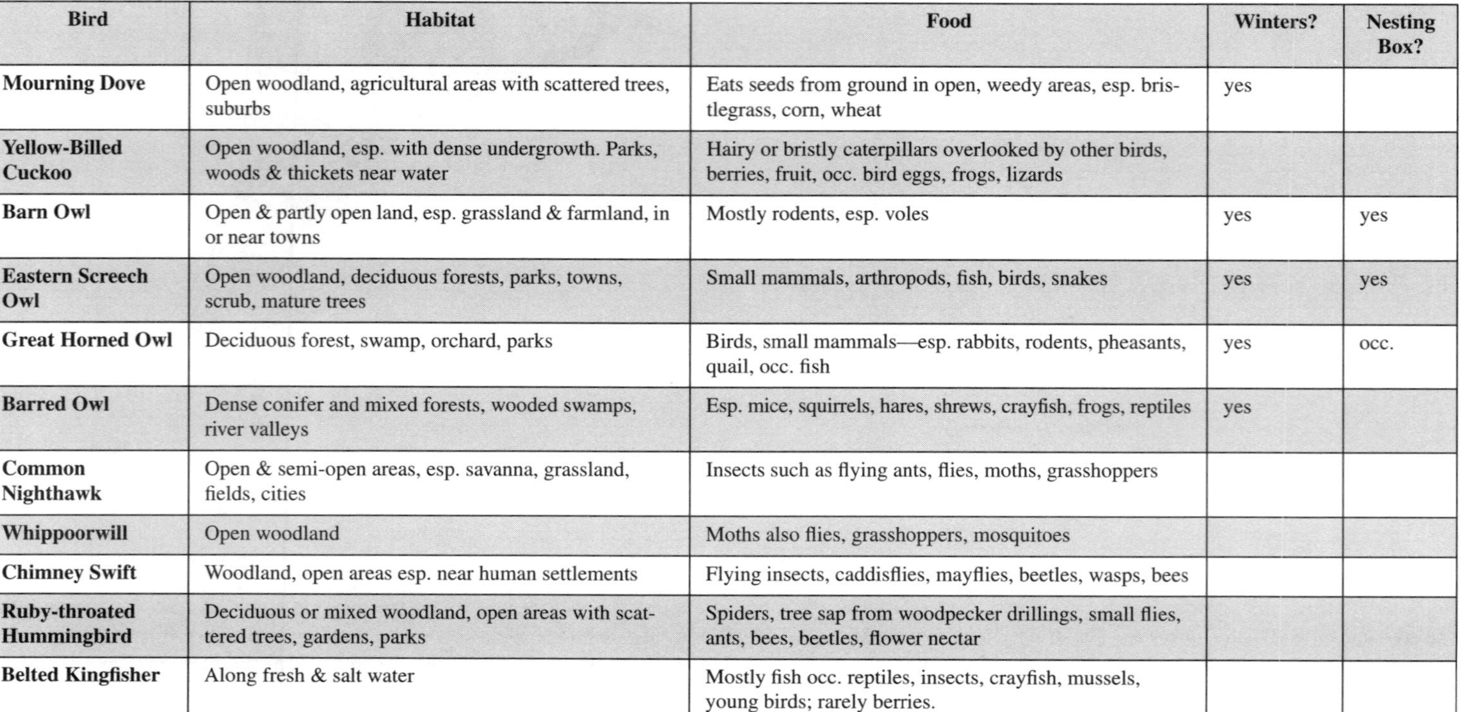

Bird	Habitat	Food	Winters?	Nesting Box?
Mourning Dove	Open woodland, agricultural areas with scattered trees, suburbs	Eats seeds from ground in open, weedy areas, esp. bristlegrass, corn, wheat	yes	
Yellow-Billed Cuckoo	Open woodland, esp. with dense undergrowth. Parks, woods & thickets near water	Hairy or bristly caterpillars overlooked by other birds, berries, fruit, occ. bird eggs, frogs, lizards		
Barn Owl	Open & partly open land, esp. grassland & farmland, in or near towns	Mostly rodents, esp. voles	yes	yes
Eastern Screech Owl	Open woodland, deciduous forests, parks, towns, scrub, mature trees	Small mammals, arthropods, fish, birds, snakes	yes	yes
Great Horned Owl	Deciduous forest, swamp, orchard, parks	Birds, small mammals—esp. rabbits, rodents, pheasants, quail, occ. fish	yes	occ.
Barred Owl	Dense conifer and mixed forests, wooded swamps, river valleys	Esp. mice, squirrels, hares, shrews, crayfish, frogs, reptiles	yes	
Common Nighthawk	Open & semi-open areas, esp. savanna, grassland, fields, cities	Insects such as flying ants, flies, moths, grasshoppers		
Whippoorwill	Open woodland	Moths also flies, grasshoppers, mosquitoes		
Chimney Swift	Woodland, open areas esp. near human settlements	Flying insects, caddisflies, mayflies, beetles, wasps, bees		
Ruby-throated Hummingbird	Deciduous or mixed woodland, open areas with scattered trees, gardens, parks	Spiders, tree sap from woodpecker drillings, small flies, ants, bees, beetles, flower nectar		
Belted Kingfisher	Along fresh & salt water	Mostly fish occ. reptiles, insects, crayfish, mussels, young birds; rarely berries.		

Bird	Habitat	Food	Winters?	Nesting Box?
Red-headed Woodpecker	Old burns, recent clearings, beech, oak, other open woodlands, open areas with scattered trees	Insects, bird eggs, nestlings, corn, berries, cherries, mulberries, seeds, acorns	yes	yes
Red-bellied Woodpecker	Swamps, parks, towns, deciduous & occ. evergreen forests	Acorns, nuts, fruits—esp. grapes; insects, corn	yes	yes
Yellow-bellied Sapsucker	Birch, poplar, aspen woodlands, often near water, mixed forests	Ants; tree sap of maple, birch, fir, hickory, beech, ash, cedar, tulip tree, magnolia, pine, oak, spruce, elm, cambium; fruit, berries		yes
Downy Woodpecker	Deciduous & mixed woodlands, parks, orchards	75-85% insects, fruit, seeds, larvae of wood-boring beetles & moths, adult beetles, ants, sap from sapsucker holes	yes	occ., to roost
Hairy Woodpecker	Deciduous or evergreen forests, wooded swamps, orchards, wooded towns or parks	75-95% insects, sap from sapsucker holes, winter: acorns, hazelnuts & beechnuts	yes	yes
Northern Flicker	Open feeding ground, snags	Ants, beetles, some grasshoppers, crickets, cockroaches, caterpillars, poison ivy berries	yes	occ.
Pileated Woodpecker	Mixed forests, open woodlands, parks, suburbs	75% insects, some fruits, acorns, tree sap, nuts; winter: dormant ants	yes	yes
Eastern Wood Peewee	Deciduous & mixed forests, forest edge, woodlands	Insects, few berries		
Least Flycatcher	Open deciduous or mixed woodlands, suburbs	Winged insects, esp. flies; berries, few seeds		
Eastern Phoebe	Open & river woodlands, rocky ravines, farmland with scattered trees	Bees, wasps, ants, occ. small fish & frogs, berries, few seeds		
Great-crested Flycatcher	Deciduous forest edge, woodlands, orchards, parks	Moths, caterpillars, berries, occ. lizards, small fruit		occ.
Eastern Kingbird	Farmland, open or river woodland, forest edge	Insects, some fruit		
Horned Lark	Open country, grassland, agricultural areas	Spiders, snails, beetles & larvae, caterpillars, grasshoppers, seeds from grass and forbs, esp. bristlegrass, ragweed		

Bird	Habitat	Food	Winters?	Nesting Box?
Purple Martin	Open country, savanna, rural lands, esp. near water	Insects		colonial
Tree Swallow	Open country, woodland edge	Flies, beetles, ants, bees, wasps; berries if insects unavailable		yes
Barn Swallow	Open country, savanna, esp. near water, agricultural areas	Insects, occ. berries		
Blue Jay	Decdiuous and mixed forests, open woodland, parks, suburbs & cities	Insects, seeds, acorns; spring: eggs & young of other birds	yes	
American Crow	Woodlands, farmlands, orchards, tidal flats	Insects, carrion, bird eggs & nestlings, seeds, esp. corn, fruit, nuts	sometimes	
Black-capped Chickadee	Deciduous or mixed woodlands, thickets, parks	Evergreen seeds, insects, including spiders; fruit; winter: eggs of moths, plant lice, katydids, spiders	yes	yes
Tufted Titmouse	Forests, woodlands, scrub, parks	Caterpillars, wasps, spiders & eggs, occ. snails; winter: acorns	yes	yes
White-breasted Nuthatch	Deciduous forests, woodlands, forest edge. Prefers mature trees	Spiders, beetles, ants, moths, caterpillars; winter: many acorns, nuts	yes	yes
Brown Creeper	Pine forests	Spiders, other insects, acorns, beechnuts, pine and corn seeds		
House Wren	Open woodlands, shrubland, agricultural land, suburbs	Millipedes, spiders, snails		yes
Marsh Wren	Marshes with many reeds	Aquatic insects, snails	in coastal areas	
Golden-crowned Kinglet	Open evergreen forests	Spiders, fruit, seeds. Young birds: insects, no spiders		
Ruby-crowned Kinglet	Evergreen and mixed forests	Insects; 10% of fall & winter food is plants		
Eastern Bluebird	Forest edge, burned or cutover woodlands, open country with scattered trees	Earthworms, snails, insects		yes

Bird	Habitat	Food	Winters?	Nesting Box?
Veery	Shaded, moist woods with understory, esp. poplar, aspen	Insects, including spiders; some fruit, esp. bunchberry, in spring and fall		
Wood Thrush	Deciduous or mixed forest, esp. near water; occ. found near suburbs or cities	Insects including spiders, caterpillars, grasshoppers; fruit often more than 33% of diet, esp. spicebush and dogwood		
American Robin	Ubiquitous	Earthworms, snails, insects, much fruit, esp. wild & cultivated cherry, dogwood, sumac		
Gray Catbird	Thick brush, often bordering water, wood suburbs, forest edge	Insects including spiders, ants, beetles, caterpillars, grasshoppers; berries, esp. blackberry, cherry		
Northern Mockingbird	Many open & partly wooded habitats, suburbs and cities	Crayfish, sowbugs, snails, berries, beetles, ants, bees, wasps, grasshoppers	most are resident	
Brown Thrasher	Brush, shrubland, deciduous forest edge & clearings	Small animals such as lizards, salamanders, frogs; berries, esp. blackberry & wild cherry	yes	
Cedar Waxwing	Woodland, forest edge, wooded suburbs	Berries, flowers, esp. red cedar berries, wild cherry, flowering dogwood; 70% of diet insects		
European Starling	Deep forest is the only habitat it avoids	Insects, seeds, berries, esp. wild and cultivated cherry, sumac	yes	
Warbling Vireo	Open deciduous & mixed woodland, forests near water, thickets	Some spiders, a few berries, caterpillars		
Red-eyed Vireo	Deciduous forests, woodland, wooded suburbs	Caterpillars, moths, beetles, ants, wasps, bees, spiders		
Tennessee Warbler	Bogs, swamps, evergreen & mixed forest edge	Insects, some berries		
Nashville Warbler	Deciduous & evergreen woodlands & woodlands near water, bogs, thickets	Insects		
Yellow Warbler	Moist, second growth woodland, gardens, scrub	Insects, few berries		
Chestnut-sided Warbler	Brushy thickets, open deciduous woodlands & borders, early second growth	Berries when insects are scarce		

Bird	Habitat	Food	Winters?	Nesting Box?
Magnolia Warbler	Open spruce/fir/hemlock forest	Insects, almost no fruit		
Yellow-rumped Warbler	Evergreen & mixed forests	Insects, berries from shrubs	some	
Black-and-white Warbler	Deciduous & mixed forests, esp. on hillsides & in ravines	Early spring: insects including dormant insects; earliest to return		
American Redstart	Open deciduous & mixed woodland, forest edge, second growth	Insects, rarely seeds, berries		
Ovenbird	Deciduous forests, rarely pine forests; evergreen forests during outbreaks of spruce budworm	Insects; worms, spiders, snails		
Northern Water Thrush	Wooded swamps, forests with standing or slow-moving water	Water & land insects, mollusks, crustaceans, occ. small fish		
Common Yellow throat Warbler	Overgrown fields, hedgerows, woodland margin, marshes	Insects, incuding spiders; few seeds		
Scarlet Tanager	Deciduous forests and woodlands, mixed forest	Wasps, bees, ants, beetles, caterpillars, moths, fruit, grains		
Northern Cardinal	Thickets, dense shrubs, undergrowth, residential areas	Insects, fruits, seeds	yes	
Rose-breasted Grosbeak	Deciduous forests, woodlands, second growth	Beetles make up nearly 50% of animal diet; seeds, fruits esp. elderberry and wild cherry; buds, some flowers, esp. wild cherry		
Indigo Bunting	Deciduous forest edge & clearings, open woodlands, weedy fields, shrublands, orchards	Insects, esp. caterpillars, grasshoppers, beetles, grain, berries		
Rufous-sided Towhee	Forest edge, thickets near water, woodlands	Insects, grass & forb seeds, acorns, berries	in southern part of range	
American Tree Sparrow	Open areas with scattered trees, brush	90% of food is weed seeds, esp. bristlegrass, crabgrass, panicgrass and sedge. Insects, few spiders, willow & birch buds & catkins, few berries.	yes	

Bird	Habitat	Food	Winters?	Nesting Box?
Chipping Sparrow	Open evergreen forest edge, oak & pine-oak woodlands, thickets, parks	Insects, few spiders, grass & forb seed		
Field Sparrow	Old fields, brush, deciduous forest edge, thorn scrub	Weed seeds main food, esp. bristlegrass, crabgrass, broomsedge and panicgrass; insects, few spiders	yes	
Fox Sparrow	Evergreen or deciduous forests, undergrowth, edge, woodland thickets, scrub, woodlands near water, mountain brushland. Feed along field edges	Main food: weed seeds, esp. smartweed, ragweed; also hawthorn & blackberry fruits; insects, few spiders, millipedes, buds		
Song Sparrow	Dense vegetation near water, forest edge, clearings, bogs, gardens	Insects, grass & forb seeds, esp. smartweed, bristlegrass, ragweed & panicgrass; some berries	yes	
White-throated Sparrow	Evergreen & mixed forests, edge and clearings, thickets, open woodlands	Insects, few spiders, millipedes, snails; forb, tree and grass seeds, such as ragwood & smartweed		
House Sparrow	Agricultural fields, woodlands & edge, cities & suburbs	Insects, including spiders; grass & forb seeds, blossoms	yes	
Dark-eyed Junco	Evergreen & deciduous forests, forest edge, open woodlands, bogs	Caterpillars, beetles, ants, few spiders, many different seeds, esp. ragweed, bristlegrass, crabgrass, dropseedgrass	yes (sometimes called "snowbird")	
Bobolink	Tall grass, flooded meadows, prairie, grain fields	Insects, few spiders, grass and forb seeds, esp. wild rice, bristlegrass		
Red-winged Blackbird	Marshes, areas near water, fields	Insects, few spiders, grass and forb seeds, esp, ragweed, bristlegrass, corn, oats, wild rice, smartweed; rarely fruit		
Eastern Meadowlark	Grassland, savanna, fields	Grasshoppers, crickets, few spiders, grass & forb seeds, some fruit		
Common Grackle	Open areas with scattered trees, open woodland, suburbs & cities	Insects, crustaceans, fish, bird eggs, nestlings, seeds, esp. corn	yes	
Brown-headed Cowbird	Woodland, forest *(esp. deciduous)*, forest edge, grassland	Grasshoppers, spiders, few snails, grain, grass & forb seeds, esp. bristlegrass, ragweed, oats & corn		

Bird	Habitat	Food	Winters?	Nesting Box?
Baltimore Oriole	Open woodlands, woodlands near water, deciduous forest edge, open areas with scattered trees, suburbs & cities	Crustaceans, few spiders, snails, nectar, few buds in spring		
Purple Finch	Open evergreen & mixed forests, forest edge, open woodlands	Seeds, some tree buds & blossoms, esp. buds & seeds of elm, fruit of tulip tree in fall & winter, apple buds in spring, cherry & peach buds & fruit, pear buds, red cedar berries	yes	
House Finch	Open woodland, urban areas, agricultural land	Fruit, buds, tree sap; almost no insects	yes	
American Goldfinch	Weedy & cultivated fields, open deciduous woodlands & woodlands near water	Seeds of deciduous trees & forbs, esp. thistles, grass seeds, flower buds, berries; caterpillars, small amount of aphids	yes	

Sources: 35, 72, and 113

Common Grackle
(Patricia Gilbert)

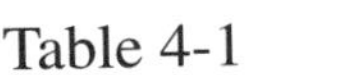

Table 4-1

TREES FOR WILDLIFE

* Identified as being at least potentially invasive in some states. Check with your state natural resources department before planting.

State	Tree/Shrub	Maximum Height	Site Preference	Animals Preferring	Used For	Makes Food Available
ME, MD, MA (higher elevations), MI, MN, WV	Arborvitae, eastern/ Northern white cedar *(thuja occidentalis)*	40-60 ft.	moist, wet	browse: deer; branches: winter cover for songbirds	cover	winter
DE, IL, IN, MD, MI, MN, NJ, OH, PA, WV, WI	Ashes *(fraxinus spp.)*	70-100 ft.	full sun, moist to wet soil	seeds for birds, mammals	food	fall
DE, MD, MI, MN, OH, PA, WV	Aspen, big-toothed *(populus grandidentata)*	70 ft.	all but constantly wet sites	branches: cover and nesting sites for woodpeckers, yellow-bellied sapsuckers; buds and catkins: grouse	cover, nesting, food	all seasons
MD, MI, MN, OH, PA, WV	Aspen, quaking *(populus tremuloides)*	50 ft.	all but constantly wet sites	buds and catkins: grouse, food of choice for beavers	food	all seasons
DE, IL(S), MD, MI, NJ, OH, PA, VT (mountains), WV	Beech, American *(fagus grandifolia)*	100 ft.	rich woods, moist areas	flickers, woodpeckers, grouse, tufted titmouse, many other birds	food	spring, fall, winter
MI, MN, PA	Birch, white *(betula papyrifera)*	70 ft.	all sites	catkins, buds: grouse; seeds: grouse, pine siskins, goldfinches, other small birds	food	fall, winter, spring
MI, MN, PA, VT (mountains), WV	Birch, yellow *(betula lutea)*	80 ft.	moist, well-drained soil, full sun or part shade	seeds: many birds; branches: red-shouldered hawk	food, perching	catkins: spring

State	Tree/Shrub	Maximum Height	Site Preference	Animals Preferring	Used For	Makes Food Available
DE, IL(S.), IN, MD, MI, NJ, OH, PA, WV	Black gum, Tupelo *(nyssa sylvatica)*	85 ft.	moist areas	purple berries for birds, esp. robins, pileated woodpeckers; deer	food	fall
MD, MN, OH, PA, WV	Buckeye *(aesculus spp.)*	55-90 ft.	rich woods	hummingbirds, butterflies, fox squirrels	food, cover, hummingbird nectar, butter-fly nectar	
IL, MI, MN, OH, PA, WV	Butternut *(juglans cinerea)*	90 ft.	bottomlands	songbirds, squirrels	food and cover	
DE, IL	Buttonbush *(cephalanthus occidentalis)*	4-15 ft.	streams, lakeshores, ponds, swamps	seed: various ducks; nectar: ruby-throated hummingbirds; branches: red-winged black-bird, Virginia rail	food, nesting	summer
DE, IL, IA, IN, MD, MA (higher elevations), ME, MI, MN, NJ, OH, PA, WV	*Cedar, eastern red *(juniperus virginiana)*	40 ft.	can grow almost anywhere	berries: songbirds, esp. evening grosbeaks, cedar waxwings	food, cover	spring, fall, winter
DE, IL, IN, MD, MI, NJ, OH, PA, WV	Cherry, black *(prunus sero-tina)*	75 ft.	moist to dry soil	white flowers, berries for birds, esp. evening grosbeaks, robins, starlings, cedar wax-wings; deer	food	summer, fall
IA, MA (higher elevations), MN, PA, WV	Chokecherry *(prunus virgini-ana)*	20-30 ft.	well-drained soil	spreads by suckering to form dense thickets; fruit: game and songbirds, esp. bluebirds, mockingbirds, catbirds	cover, food	summer, fall, winter

State	Tree/Shrub	Maximum Height	Site Preference	Animals Preferring	Used For	Makes Food Available
IL, MA (higher elevations), MN, NJ, PA, WV	Crabapple *(malus spp)*	25 ft.	thickets in moist areas	29 species of bird use fruit, seeds or buds	food, nesting, cover	winter
IA, WI (south)	Sargent crab *(malus sargenti)*	15-30 ft.	best on well-drained loam, prefers full sun, tolerates variety of sites	deer eat fruit		
IA, MA (higher elevations), MN, PA, VT, WI	Cranberry, American high-bush *(viburnum trilobum)*	10-13 ft.	well-drained to moist	berries persist through winter grouse, songbirds and squir-rels use for emergency food	food	winter
NJ, PA, MD, WV	Blackhaw viburnum *(viburnum prunifolium)*	30 ft.				
IN	Cyprus, bald *(Taxodium distichum)*	to 100 ft.	clay or silty soil, wet to moist soil	nesting, cover	cover	
DE, IL(S&C), IN, MD, MA (higher elevations), MI, NJ, OH, PA, WV	Dogwood, flowering *(cornus florida)*	30 ft.	full to partial sun, rocky woods, wooded slopes	white flowers, red berries for birds, esp. cardinals, evening grosbeaks, robins, wood thrushes, cedar waxwings; rab-bits, squirrels, eastern chipmunks	food	leaves, twigs, fruit: fall, winter flowers: fall, summer
IA, MN, VT, WI	*Dogwood, gray *(cornus racemosa)*	7-10 ft.	moderately shade tolerant, does best in full sun, well-drained silt loam	for cover: woodcock, song-birds; fruit: songbirds, small mammals, deer	food, cover	leaves, twigs, fruit: fall,winter flowers: fall. summer
DE, IL(N&C), MN, VT	Dogwood, pagoda *(cornus alternifolia)*	20 ft. or less	sun to shade, rich, moist soils	fruits: ruffed grouse, song-birds, esp. cardinals, robins, evening grosbeaks; rabbits, squirrels, eastern chipmunks	food	leaves, twigs, fruit: fall, winter flowers: fall, summer

State	Tree/Shrub	Maximum Height	Site Preference	Animals Preferring	Used For	Makes Food Available
IL, IA, MN, VT, WI	*Dogwood, red osier *(cornus sericea)*	10-12 ft.	wet to well-drained	fruits: turkey, grouse, woodpeckers, quail, songbirds, wood ducks; browse: deer	food	leaves, twigs, fruit: fall, winter flowers: fall, summer
DE, IA, MN, VT, WI	Dogwood, silky *(cornus amonum)*	4-10 ft.	moist to well-drained, sun or shade	all plant parts: many birds and mammals	food, cover	leaves, twigs, fruit: fall, winter flowers: fall, summer
DE, IL, IA, MN, VT, WV	Elderberry *(sambucus spp.)*	5-13 ft.	moist to well-drained soil, sun or shade	29 species of birds and mammals, incl. orioles	food, cover	fall, winter, early spring
ME, MI, MN, NH, VT, WV	Fir, balsam *(abies balsamea)*	40-60 ft.	cool sections of north and higher elevations (NH)	seeds: chickadees, red-breasted nuthatches, evening grosbeaks, other birds	food, roosting, nesting, cover	fall, winter
IL, IA, IN, MD, MI, MN, OH, PA, WV	Hackberry *(celtis occidentalis)*	40-50 ft.	tolerates droughty or alkaline soils, shade and wet to dry sites	seeds: wood ducks, cardinals, grosbeaks, mockingbirds, robins, songbirds; deer, small mammals	food	
DE, IN, MN, NJ, OH, VT, WV	Hawthorns *(Crataegus spp.)*	20-30 ft.	dry, moist	food: birds and small mammals; cavity nesting birds in older tree trunks	food, nesting	fall
MN, WI	Hazelnut, American *(corylus americana)*	8 ft.	variety of loams	berries: esp. red-bellied woodpeckers; cover for ruffed grouse, songbirds	food, cover	fall, winter, spring
ME, MD, MI, MN, NJ, OH, PA, VT, WV	Hemlock, eastern *(Tsuga canadensis)*	60-70 ft.	will grow in shade, grows well near streams	nuts: deer, squirrels, jays, hairy woodpeckers, pheasants catkins: ruffed grouse	food	winter

State	Tree/Shrub	Maximum Height	Site Preference	Animals Preferring	Used For	Makes Food Available
DE, IL,IN, MD, MI, MN, OH, PA, WV	Hickory, shagbark *(Carya ovata)*	80 ft.	full to partial sun	branches: wood thrushes, robins, other birds; seeds: chickadees, pine siskins, goldfinches	nesting, food	fall, winter
IL, MD, MI, MN, VT, WV	Hophornbeam *(ostrya virginiana)*	less than 50 ft.	upland woods, slopes near streams	seeds for squirrels, songbirds	food	
IL, IA, MD, MI, MN, PA, WV	Linden, American basswood *(tilia americana)*	30-50 ft.	moist, well-drained. Can grow in shade but does best in full or partial sun	seeds: songbirds; den trees for small mammals	food, nesting	
DE, IL, IN, MD, MI, MN, NJ, OH, PA, VT, WV	Maples *(Acer spp.)* *Norway, Amur	65-100 ft.	Woodlands — will shade out oaks	songbirds, small mammals	food, cover	flowers, fruit: fall leaves; twigs: spring, winter; bark: fall, winter, spring
MA (higher elevations), MI, WV **MN** **MD, PA**	Mountain ash, American *(sorbus americana)* Mountain ash, showy *(sorbus decora)* Mountain ash, European *(sorbus aucuparia)*	20-40 ft.	moist to dry soil, cool climate	birds	food	fall, winter
DE, IL, IA, MD, MI, MN, PA, WV (rare)	Mulberry, red *(morus rubra)*	25-40 ft.	sun to light shade, moist to well-drained soil; fruit leaves purple stains—don't plant near buildings	small birds, mammals	food, cavities for nesting birds	spring, summer
IA, MN, PA	Nannyberry *(viburnum lentago)*	15-25 ft.	Dry, well-drained or moist soil	birds	food	winter, spring

State	Tree/Shrub	Maximum Height	Site Preference	Animals Preferring	Used For	Makes Food Available
IL, IA, MN, WI	Ninebark *(physocarpus opulifolius)*	10 ft.	will grow on droughty sites	songbirds	cover, food	
DE, IA, IN, MD, MN, NJ, OH, PA, VT, WV, WI	Oaks *(quercus spp.)*	60-100 ft.	mildly acidic to neutral soil, full sunlight, well-drained sites	buds: ruffed grouse; seeds: songbirds	food, cover-when planted in several rows	fall
DE, IN, MD, WV	Pawpaw *(Asimina triloba)*	30 ft.	partial sun, often grows in clusters	acorns for most birds and mammals	food	
ME, MD, MI, MN, PA	Pine, red *(pinus resinosa)*	50-75 ft.	tolerates poorer soil	birds	food	fall, winter
IN, ME, MD, MI, MN, NJ, OH, PA, VT, WV	Pine, white *(pinus strobus)*	80-100 ft.	full sun	pine siskin	food	
IL, IA, WI **DE, IN**	Plum, wild *(prunus spp.)* Plum, American, *(prunus americana)*	15 ft. or more	full sunlight, well-drained silt loam, full to partial sun	birds, esp. robins, mourning doves, blue jays, upland ground birds, small mammals, deer	food, nesting, cover	summer
DE, IL(S), IN, MD, MI, NJ, OH, PA, WV	Poplar, tulip *(liriodendron tulipifera)*	150 ft. (one of the largest of eastern hardwoods)	rich woods	hummingbirds, honey bees, songbirds	nectar. food	fall, winter, spring
IL, IN, MD, MI, PA, WV	Redbud, eastern *(cercis canadensis)*	50 ft.	rich woods, ravines, fencerows	fruit: songbirds; thickets from root sprouts	food, nesting cover	
DE, IA, MD, NJ, PA, VT, WV **IL, MN**	Serviceberry, shadbush *(amelanchier canadensis)* Allegheny serviceberry *(amelanchier arborea)*	20-30 ft.	well-drained, moist soil	purple flowers, seeds for birds, incl. orioles, deer	food	summer

State	Tree/Shrub	Maximum Height	Site Preference	Animals Preferring	Used For	Makes Food Available
DE, IL	Spicebush *(lindera benzoin)*	15 ft.	partial shade, moist, well-drained soil, acidic soil	fruits: wood thrush, veery many stems yield good screening	food, cover	
MI, MN, NH, VT (N.E. Highlands)	Spruce, black *(picea mariana)*	30-70 ft.	cool sections of north and higher elevations (NH)	twigs: deer, rabbits, opossums; fruits: ruffed grouse, songbirds, esp. thrushes	food	winter
ME (exotic), MD, MI, WI (central, south), WV	Spruce, Norway *(picea abies)*	100 ft. +	wide variety, prefers well-drained silt loam		cover	winter
ME, NH, VT, WV	Spruce, red *(picea rubens)*	60-70 ft.	cool sections of north and higher elevations (NH)		food, cover	winter
ME, MI, MN, NH, WI, VT (n. central)	Spruce, white *(picea glauca)*	60-70 ft.	cool, moist sites	cones: red and gray squirrels	food, cover—plant 50 or more	winter
IL, MD, MN	*Sumac, smooth *(rhus glabra)*	20 ft.	upland soil, mesic to dry	birds, deer, small mammals	food	fall, winter
IL(N&C), MD, MI, MN, WV	Sumac, staghorn *(rhus typhina)*	35 ft.	dry soil, sandy ridges	birds, small mammals	food, escape cover	fall, winter
ME, MI, MN, NH, PA, WV	Tamarack/ Eastern larch *(larix laricina)*	40-75 ft.	cool sections of north and higher elevations (NH), prefers wet areas (WV)	fruit: robins, thrushes, catbirds, thrashers, others	food during spring and fall migrations	fall
WI	Thornapple *(crataegus columbiana)*	20-24 ft.	prefers full sun, silt loams	seedlings, saplings: snowshoe hare, spruce grouse, small mammals, mature cone seeds: red squirrels, pine grosbeaks, crossbills, purple finches	food, also good nest tree	

State	Tree/Shrub	Maximum Height	Site Preference	Animals Preferring	Used For	Makes Food Available
DE, IN, MD, MN, OH, PA	Willow *(Salix spp.)*	50ft. +	moist, wet	small mammals like squirrels	food	leaves, bark: spring, summer
MN	Pussy willow *(salix discolor)*	10 ft.				
DE, MN **WV**	Winterberry *(ilex verticillata)* *(ilex montana)*	10 ft.	fruit persists through winter	buds: grouse; twigs: goldfinches	food, nesting cover	fruit: fall, winter
DE, MD, NJ, PA	American holly *(ilex opaca)*	to 50 ft.				
ME, MN	Yew, American *(taxus canadensis)*	maximum: 3 ft.	prefers damp, mucky soil and shade of other evergreens	songbirds	food	

Note: The trees in this chart were mentioned in the following publications. Generally, they are trees native to the states where they are listed. Soil type, amount of sunshine and exposure of the site to be planted should also be taken into account. Check with your area forester before choosing trees to plant.

Sources : 48 (IL), 85 (IN), 53 (IA), 66 (MD), personal correspondence (DE, MA), 47 (ME), 61 (MI), 77 (MN), 130 (NH), 80 (NJ), 31 (OH), 33 (PA), 90 (PA), 59 (VT), 98 (VT), 6 (VT), 119 (WV), 122 (WI), 123 (WI). Also 89, 115.

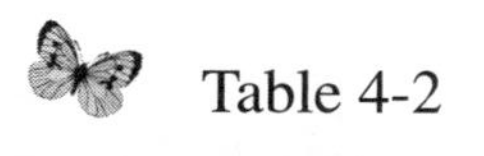

Table 4-2

PREFERRED PLANTS OF SOME COMMON BUTTERFLIES

Butterfly	Food For Larvae	Nectar Plants
American Lady	Sunflower, pearly everlasting, holly-hock, common mallow, many others	Joe-pye weed, aster, zinnia, goldenrod, butterfly bush, buttonbush, monarda, sedum, privet, wallflower, globe thistle, anise hyssop, pincushion flower, phlox, passion flower
Buckeye	Snapdragon	aster, milkweed, chickory, coreopsis
Cabbage White	Cabbage, wallflower, nasturtium, cleome and other mustards and capers	mints, aster, dandelion, clover, selfheal, monarda, passion flower, wallflower, pincushion flower, lantana, dame's rocket and other mustards
Clouded Sulphur	Alfalfa, clovers, vetch, trefoil, other legumes	clover, tithonia, phlox, milkweed, goldenrod, dandelion, aster, dame's rocket, pincushion flower, globe thistle, marigold, scarlet sage, sedum, zinnia
Comma	Nettle, elm, hackberry, false nettle	tree sap, butterfly bush, dandelion, asters, milkweed, zinnia
Eastern Tailed Blue	Beggar's tick, vetch, alfalfa	milkweed, cinquefoil, wild strawberry, fleabane, coreopsis, zinnia, wild geranium
Giant Swallowtail	Rue, prickly ash, Hercules' club	lantana, butterfly bush, goldenrod, purple coneflower, Joe-pye weed, blanket flower, phlox, dame's rocket, scarlet sage, lilac, milkweed, azalea
Great Spangled Fritillary	Violet	coneflower, Joe-pye weed, ironweed, milkweed, black-eyed susan, verbena, blanket flower, mountain laurel, monarda, red clover, globe thistle, lantana, catmint, scarlet sage, buttonbush
Great Swallowtail	Prickly ash	lantana, milkweed, lilac, goldenrod, azalea
Milbert's Tortoiseshell	Nettles	sneezeweed, goldenrod, aster, marigold, daisy, wallflower, ageratum, butterfly bush, lilac, sedum, rock cress, monarda, Joe-pye weed, zinnia, anise hyssop, tree sap
Monarch	Milkweed	milkweed, butterfly bush, goldenrod, thistle, ironweed, mints, aster, Joe-pye weed, blanket flower, gay-feather, cosmos, lantana, scarlet sage, abelia, lilac, mallow, sedum, zinnia

Butterfly	Food For Larvae	Nectar Plants
Mourning Cloak	Willow, elm, poplar, aspen, birch, hackberry, cottonwood	tree sap, butterfly bush, milkweed, shasta daisy, dogbane, rock cress, moss pink, zinnia
Painted Lady	Daisy, hollyhock	goldenrod, aster, zinnia, butterfly bush, milkweed
Pearl Crescent	Asters	asters, daisy, black-eyed Susan, dogbane, milkweed, geranium, wallflower, mint, zinnia, coreopsis
Red Admiral	Nettle	tree sap, daisy, aster, goldenrod, butterfly bush, milkweed
Spring Azure	Dogwood, blueberry, viburnum, cherry, sumac, privet	dogwood, holly, blackberry, milkweed, spicebush, forget-me-not, dandelion, privet, lilac, rock cress, cherry, violet, coreopsis
Tiger Swallowtail	Cherry, ash, birch, tulip tree, lilac, aspen, hornbeam, willow, spicebush	butterfly bush, milkweed, phlox, lilac, ironweed, monarda, buttonbush, lantana, Hercules' club
Viceroy	Willow, poplar, apple, aspen, plum, cherry	tree sap, Joe-pye weed, catmint, phlox, aster, goldenrod, milkweed

Sources: 3, 69

TABLE 4-3

PLANTS FOR BIRD AND BUTTERFLY GARDENS

ANNUAL FLOWERS

Annual	Color	Height	Makes Available	Attracts
Bachelor's Button, Cornflower *(Centaurea Cyanus)*	blue, pink	1-2 ft.	nectar	B
Canna *(Canna X Generalis)*	red, orange, yellow	3-6 ft.	nectar	H
Cosmos *(Cosmos Bipnnatus)*	red, pink, crimson, white	3-4 ft.	nectar	B
Dahlia	yellow, red, pink, violet	1 ft. or taller	nectar	H
Flowering Tobacco *(Nicotiana)*	fragrant trumpets crimson, rose pink, white	1-3 ft.	nectar, seed	H, F&J, B
Four O'clocks *(Miriabilis Jalapa)*	rose, white	2 ft.	nectar, seeds	H, F&J, B
Fuchsia *(Fuchsia Riccartoni)*	rose and purple, white and rose	2 ft.	nectar	H
Gladiolus	red, pink, purple, white	2-4 ft.	nectar	H
Heliotrope *(Heliotropium Arborescens)*	deep purple	2 ft.	nectar	B
Larkspur *(Consolida)*	blue, violet	12-18 inches	nectar	H
Marigolds *(Targetes Spp.)*	red, orange, yellow	8 in. to 3 ft.	nectar, larvae food	H, B, BL
Mexican Sunflowers *(Tithonia Rotundifolia)*	orange-yellow	4-6 ft.	nectar	B
Nasturtium *(Tropaeolum Majus)*	red, orange, yellow	low-growing or climbing	nectar, aphids (hummingbird food)	H

Annual	Color	Height	Makes Available	Attracts
Pansy *(Viola)*	blue, red, yellow, white	8-12 inches	nectar, larvae food	B, BL
Petunia	red, deep pink, purple, blue, white	8-18 inch	nectar, seeds	H, F&J, B
Pincushion Flower ***(Scabiosa Atropurpurea)***	dark purple, pink, white	2-3 ft.	nectar	H
Pot Marigold ***(Calendula Officinalis)***	yellow, orange	2 ft.	nectar	B
Sage ***(Salvia Splendens)***	scarlet	1-3 ft.	nectar, seeds	H, F&J
Spider Flower, Cleome ***(Cleome Hasselerana)***	pink	5 ft.	nectar, seeds	H, F&J, B
Sunflower ***(Helianthus Annuus)***	yellow, black center	to 12 ft.	nectar, seeds	S, B
Sweet Alyssum ***(Lobularia Maritima)***	white	1 ft.	nectar, larvae food	B, BL
Sweet Pea ***(Lathyrus Odoratus)***	white, pink, rose	vines to 8 ft.	nectar, larvae food	B, BL
Sweet William ***(Dianthus Barbatus)***	white, pink, rose, purple	1-1/2 ft.	nectar	H, B
Touch-me-not, Jewelweed ***(Impatiens Spp.)***	yellow	3 ft.	nectar, seeds	H, F&J
Verbena ***(Verbena X Hybrida)***	fragrant red, pink, yellow, purple, white	trailing or upright, usually under 1 ft.	nectar	H
Wax Begonia ***(Begonia X Semperflorens-Cultorum)***	white, pink, red	8 inches to 1 ft.	nectar	H
Zinnia ***(Zinnia Elegans)***	red, rose, yellow, white	8 in. to 3 ft.	nectar, seeds	H, F&J, B

PERENNIAL FLOWERS

Perennial	Color	Height	Hardiness Zone	Time of Bloom	Makes Available	Wildlife Attracted
Aster	New England aster *(a. novae-angliae)* and New York aster *(a. novae-beglii)*: purple, white, pink petals, yellow center	3-5 ft.	4, 5, 6	August-Sept.	nectar, larvae food	B, BL
Astilbe	purple, pink, white	2-3 ft.	4, 5, 6	June	nectar	B
Bee Balm, Bergamot, Oswego Tea *(Monarda Didyma)*	pink, red, light purple	2-4 ft.	4, 5, 6	June-August	nectar	H, B
Canterbury Bell, Bellflower *(Campanula Medium)*	blue, white	3 ft.	4, 5, 6	May-June	nectar, seeds	J&F, B
Black-eyed Susan *(Rudbeckia Spp.)*	yellow	1-3 ft.	4, 5, 6	July-August	nectar, larvae food	B, BL
Blanket Flower *(Gaillardia)*	red and yellow	1-3 ft.	4, 5, 6	June, July	nectar	B
Butterfly Bush *(Buddleia Davidi)*	purple	4-10 ft.	4, 5, 6	July (late) to frost	nectar	H, B
Butterflyweed *(Asclepias Tuberosa)*	orange	1-3 ft.	4, 5, 6	August-Sept.	nectar, larvae food	B, BL
Cardinal Flower *(Lobelia Cardinalis)*	scarlet	2-4 ft.	4, 5, 6	July-Sept.	nectar	H, B
Columbine *(Aquilegia Spp.)*	red & yellow *(a. canadensis)* or hybrids with red or pink flowers	1-2 ft.	4, 5, 6	May	nectar	H
Coralbells *(Heuchera Sanguinea)*	red	2-1/2 ft.	4, 5, 6	June-August	nectar	H
Coreopsis	yellow	12-18 inches	4, 5, 6	June	nectar	B

Perennial	Color	Height	Hardiness Zone	Time of Bloom	Makes Available	Wildlife Attracted
Dame's Rocket ***(Hesperis Matronalis)***	rose	4 ft.	4, 5, 6	May	nectar, larvae food	B, BL
Daylilies ***(Hemerocallis Spp.)***	bright red, orange, pink, yellow	2 to more than 4 ft.	4, 5, 6	June-July	nectar	H, B
Delphinium ***(Delphinium)***	deep purple, blue, light blue	2-6 ft.	4, 5, 6 (may need winter protection in 4)	June	nectar	H
Evening Primrose ***(Oenothera Spp.)***	yellow	4 ft.	4, 5, 6	July	seeds	J&F
Fireweed ***(Epilobium Angustifolium)***	rose-purple	4 ft.	4, 5, 6	July	nectar	H
Fleabane ***(Erigeron Spp)***	purple, yellow, red	10-20 inches	4, 5, 6	July	nectar	B
Foxglove ***(Digitalis Spp.)***	deep pink, rose, white, yellow tubular flowers — common foxglove *(d. purpurea)* rose pink — strawberry foxglove *(D. x mertonensis)*	2-5 ft. 3-4 ft.	5, 6	June	nectar	H
Gayfeather, Blazing Star ***(Liatris Spp.)***	purple	2-5 ft.	4, 5, 6	July-Sept.	nectar, seeds	H, J&F, B
Gentian ***(Gentiana Septemfida)***	blue (dwarf)	1 ft.	4, 5, 6	August-Sept.	nectar	H
Globe Thistle ***(Echinops Spp.)***	blue	4 ft.	4, 5, 6	July-Sept.	nectar	B
Goldenrod ***(Solidago Spp.)***	yellow	3-7 ft.	4, 5, 6	August-Sept.	nectar	B
Hibiscus ***(Hibiscus)*****, Rose Mallow**	red, rose, pink or white	5-8 ft.	5-10		nectar	H, B
Hollyhock ***(Althaea Rosa)***	burgundy, red, deep pink, orang-yellow, white	6 ft.	4, 5, 6	July-August	nectar, larvae food	H, B

Perennial	Color	Height	Hardiness Zone	Time of Bloom	Makes Available	Wildlife Attracted
Hosta *(Hosta Spp.)*	purple	2 ft.	4, 5, 6	July-August	nectar	H
Joe-pye Weed *(Eupatorium Purpereum)*	white	3-6 ft.	4, 5, 6	August-Sept.	nectar	B
Lavender *(Lavandula Spp.)*	purple	1-3 ft.	4, 5, 6 (may need winter protection in 4)	July-Sept.	nectar	B
Lilies *(Lilium)*	red, orange-red, orange, yellow	2-8 ft.	4, 5, 6	June, July	nectar	H
Lupine *(Lupinus Spp.)*	blue, pink, yellow, white	flowers 2-1/2-3-1/2 ft. above 2-ft. leaf cluster	4, 5, 6	May-July	nectar, larvae food	H, B, BL
Meadow Rue *(Thalictrum Spp.)*	yellow	3 ft.	4, 5, 6	June-August	nectar	B
Milkweed *(Asclepias Spp.)*	pink	3-5 ft.	4, 5, 6	July-August	nectar, larvae food	B, BL - esp. monarchs
Mountain Bluet *(Centaurea Montana)*	blue	2 ft.	4, 5, 6	May	nectar	B
Pearly Everlasting *(Anaphalis Margaritacea)*	white	2 ft.	4, 5, 6	June-July	larvae food	BL
Peony *(Paeonia Spp.)*	pink, red, white—single flowers	3 ft.	4, 5, 6	May-June	nectar	B
Phlox *(Phlox Spp.)*	lavender, purple, pink, red, white	4 ft.	4, 5, 6	June-August	nectar, seeds	H, J&F, B

Perennial	Color	Height	Hardiness Zone	Time of Bloom	Makes Available	Wildlife Attracted
Pincushion Flower ***(Scabiosa Caucasica)***	blue, red, pink, white	1-1/2 - 2-1/2 ft.	4, 5, 6	June	nectar	H, B
Pinks ***(Dianthus Chinensis)***	red, white, lilac	12-18 inches	4, 5, 6	May	nectar	H, B
Prunella ***(Prunella Vulgaris)***	purple (violet)	2 ft.	4, 5, 6		nectar	B
Purple Coneflowers ***(Echinacea Spp.)***	purple, white	4 ft.	4, 5, 6	July	nectar	B
Queen Of The Prairie ***(Filipendula Rubra)***	pink (magenta)	4-7 ft.	4, 5, 6		nectar	B
Queen-anne's Lace ***(Daucus Carota)***	white	2 ft.	4, 5, 6	June-Sept.	nectar, larvae food	B, BL
Sedges ***(Carex Spp.)***	green (tiny)	1 ft.	4, 5, 6		larvae food	BL
Sedum ***(Sedum Spectabile)***	pink	2 ft.	4, 5, 6	August-Sept.	nectar, larvae food	B, BL
Shasta Daisy ***(Chrysanthemum Maximum)***	white	1-2 ft.	4, 5, 6	June-July	nectar	B
Snapdragon ***(Antirrhinum Majus)***	red (reddish-purple), white	1-3 ft.	4, 5, 6	May-June	nectar	H, B
Sneezeweed ***(Helenium Autumnale)***	yellow	4-6 ft.	4, 5, 6	August-Sept.	nectar	B
Sunflower ***(Helianthus Laetiflorus Maximiliani)***	yellow	8 ft. to 12 ft.	4, 5, 6	August-Oct.	nectar	B
Turk's Cap Lily ***(Lilium Michi-ganese)***	orange	6 ft.	4, 5, 6	July	nectar	H
Turtlehead ***(Chelone Glabra)***	white	2 ft.	4, 5, 6	July-August	nectar	B

Perennial	Color	Height	Hardiness Zone	Time of Bloom	Makes Available	Wildlife Attracted
Violet ***(Viola Spp.)***	purple, yellow, white	1 ft.	4, 5, 6	April	nectar, larvae food	B, BL
Wallflower ***(Cheiranthus Spp.)***	yellow, orange	1 ft.	4, 5, 6		nectar, larvae food	B, BL
Yarrow ***(Achillea Spp.)***	yellow, pink, red	2 ft.	4, 5, 6	July-Sept.	nectar	B

VINES

Vines	Color	Height	Hardiness Zone	Time Of Bloom	Makes Available	Wildlife Attracted
Bittersweet ***(Celastrus Scandens)***		20 ft.			fruit	S
Dutchman's Pipe ***(Aristolochia Durior)***	yellowish-brown, u-shaped	30 ft.	4, 5, 6	July-Sept.	larvae food	BL
Morning Glory ***(Ipomoea)***	cardinal climber (i. multifida): crimson, white eyes cypress vine (i. quamoclit) red red morning glory (i. coccinea) fragrant, red	10-12 ft.	annual	June-Sept.	nectar	H
Scarlet Runner Beans ***(Phaseolus Coccineus)***	scarlet, white	10 ft., also bush variety, 1-1/2 ft.	annual	July-August	nectar, larvae food	H, B, BL
Trumpet Vine ***(Campsis Radicans)***	orange, orange red	woody vines, 30-40 ft.	5, 6; hardy in protected areas of zone 4	July-Sept.	nectar	H
Wisteria ***(Wisteria Spp.)***	purple	20 ft.	6	June-July	nectar, larvae food	H, B, BL
Vine Honeysuckle ***(Lonicera Spp.)***	red to orange	12-20 ft.	4, 5, 6	May-June	nectar	H, B

Note: Time of bloom column should be used to compare plants in order to extend blooming periods in your garden. Exact bloom times may vary somewhat from region to region.

ANNUAL AND PERENNIAL HERBS

Note: If you expect to have any of the larvae food available for your use as well as for the caterpillars' dinners, be sure to plant enough.

Perennial	Height	Hardiness Zone	Time of Bloom	Makes Available	Attracts
Hyssop *(hyssopus officinalis)*	2 ft.	4, 5, 6		nectar	B
Thyme	1 ft.	5, 6	July-August	nectar	B

Annual	Height	Makes Available	Attracts
Dill *(anethum graveolens)*	3 ft.	larvae food	BL
Fennel *(f. vulgare)*	3-5 ft.	larvae food	BL
Parsley *(Petroselinum crispum)*	8 inches	larvae food	BL

Key to wildlife attracted:

H = hummingbirds
F&J = finches and juncoes
S = songbirds
B = butterflies
BL = butterfly larvae

Note: Annuals purchased from a greenhouse will often be blooming at the time of purchase. Planted from seed, annuals will generally begin to bloom in June to early July.

Sources: 37, 44, 108

TABLE 4-4

COMMERCIAL BIRD FOOD AND HUMAN FOOD THAT BIRDS LIKE

BIRD	FEEDER FOOD	KITCHEN FOOD
American Goldfinch	niger thistle; also sunflower chips, black oil sunflower seed, white proso millet seed	
American Tree Sparrow	white proso millet, black oil sunflower seeds—feed on ground suet	bread
Black-capped Chickadee	black oil sunflower seeds, suet; also striped sunflower seeds, safflower seeds	peanuts, peanut butter, pumpkin seeds, old bread, doughnuts
Blue Jay	striped sunflower seeds; also black oil sunflower seeds, safflower seeds, cracked corn, peanut kernels	peanuts, pumpkin seeds, broken walnuts, bread, cornbread, crackers
Bobwhite Quail	birdbath on ground, small food plots of lespedeza	
Brown Creeper	suet mix	bread, peanut butter rubbed into tree bark using upward motion
Brown Thrasher	black oil sunflower seeds, suet	bread, peanut butter, raisins, popcorn, cheese
Cardinal	black oil sunflower seeds; also striped sunflower seeds, safflower seeds, cracked corn, red and white millet, wheat mealworms	
Chipping Sparrow	red or white proso millet on ground, mealworms	
Downy Woodpecker	suet	meat scraps, cracked pecans, peanut butter, cheese, fruit
Eastern Bluebird	mealworms	currants, raisins, bread, cake, pitted dates, dried figs, peanuts, peanut butter, pecans

BIRD	FEEDER FOOD	KITCHEN FOOD
Eastern Towhee	shelled corn, cracked corn, black oil sunflower seeds—sprinkled on the ground, suet	cracker crumbs, doughnuts, peanuts, nut meats, watermelon seeds
Evening Grosbeak	black oil sunflower seeds (warning: heavy feeders on these); also striped sunflower seeds, safflower seeds, cracked corn, peanuts	
Fox Sparrow	cracked corn, white proso millet, oats, black oil sunflower seeds—feed on ground, suet	peanuts, bread scraps
Golden-crowned Kinglet	suet	
Gray Catbird	suet, mealworms	apples, bread, cake, dried currants, peanuts, cheese, raisins, grapes
Hairy Woodpecker	suet, striped sunflower seeds; also black oil sunflower seeds, peanut kernels	peanut butter, hummingbird nectar
House Finch	black oil sunflower seeds, white proso millet, cracked corn, suet	peanuts, bread
Indigo Bunting	white proso millet, thistle seed	
Junco	white proso millet, black oil sunflower seeds, cracked corn, sorghum—scattered on the ground, suet	
Mourning Dove	black oil sunflower seeds, peanut hearts, white proso millet—scatter in short grass; also milo sorghum, wheat	
Northern Flicker	suet	peanut butter, raisins, apples
Northern Mockingbird	suet, mealworms	cottage cheese, apples, currants, nutmeats, peanut butter, doughnuts, bread crumbs
Northern Oriole	meal worms, suet, 4:1 water:sugar nectar	orange halves, grape jelly, broken walnuts, apple slices, bread
Pheasant	ear corn, shell corn	

BIRD	FEEDER FOOD	KITCHEN FOOD
Pileated Woodpecker	suet in feeder firmly attached to a post	
Pine Grosbeak	striped sunflower seeds; also black oil sunflower seeds, safflower seeds, cracked corn	
Pine Siskin	black oil sunflower seeds, niger thistle seeds, white proso millet, sunflower chips; will eat from thistle feeder	
Purple Finch	safflower seeds, black oil sunflower seeds; also striped sunflower seeds, red millet	peanuts, pumpkin seeds, cornbread, broken walnuts, squash seeds, peanut butter
Red Crossbill	black oil sunflower seeds, thistle seeds, salt	
Red-bellied Woodpecker	cracked corn, shell corn, ear corn, suet	bread, peanuts, orange halves, peanut butter, cracked walnuts
Red-breasted Nuthatches	suet, black oil sunflower seeds	peanut butter, cantaloupe seeds, doughnuts, nut meats
Red-headed Woodpecker	suet, striped sunflower seeds, corn; also black oil sunflower seeds, peanut kernels	cracked walnuts, bread, peanuts, orange halves (in summer)
Redpoll	niger thistle mixed with peanut hearts & sunflower chips in tube feeder with holes no more than 1/4" in diameter	
Robin	mealworms	apples, bread, toast, cornbread, figs, prunes, raisins, nutmeats, popcorn, peanuts, peanut butter, grape jelly
Rose-breasted Grosbeak	black oil sunflower seeds, suet	
Ruby-crowned Kinglet	suet, might take nectar from hummingbird feeder	cracked nuts, peanuts
Ruby-throated Hummingbird	4:1 water:sugar nectar	attract fruit flies by hanging a small bag with over-ripe bananas or cantaloupe near hummingbird feeder
Song Sparrow	cracked corn, white proso millet, black oil sunflower seeds	walnut meats, peanut butter to feed young

BIRD	FEEDER FOOD	KITCHEN FOOD
Tufted Titmouse	striped sunflower seeds; also black oil sunflower seeds, safflower seeds, red and white millet	nuts, cantaloupe, bread, cookie crumbs, cornbread, peanut butter, pie crusts
White-breasted Nuthatch	suet, black oil sunflower seeds, safflower seeds, sugar water in oriole feeder	peanuts
White-crowned Sparrow	black oil sunflower seeds, white proso millet, cracked corn—feed on ground	
White-throated Sparrow	white proso millet, cracked corn—feed on ground	
Wild Turkey	corn, wheat, black oil sunflower seeds—scattered on the ground	
Yellow-bellied Sapsucker	suet, 4:1 water:sugar nectar	peanut butter, cracked walnuts, fruits
Yellow-rumped Warbler	suet	broken walnuts, almonds, bread, cake, cashew nuts, cornbread, peanuts, peanut butter, dried figs

Sources: 45, 95

Gray Catbird
(Patricia Gilbert)

TABLE 4-5

NEST BOX DIMENSIONS

Bird	Box Depth	Box Width	Box Height	Entrance Height	Portal Diameter	Height Above Ground
Bluebird	5"	5"	8-12"	6"	1-1/2"	4-6 ft.
Chickadee	4"	4"	8-10"	6-8"	1-1/8"	4-15 ft.
Titmouse	4"	4"	10-12"	6-10"	1-1/4"	5-15 ft.
Great Crested Flycatcher	6"	6"	8-12"	6-10"	1-3/4"	5-15 ft.
Red-breasted Nuthatch	4"	4"	8-10"	6-8"	1-1/4"	5-15 ft.
White-breasted Nuthatch	4"	4"	8-10"	6-8"	1-3/8"	5-15 ft.
Downy Woodpecker	4"	4"	8-10"	6-8"	1-1/4"	5-15 ft.
Hairy Woodpecker	6"	6"	12-15"	9-12"	1-1/2"	8-20 ft.
Northern Flicker	7"	7"	16-18"	14-16"	2-1/2"	6-20 ft.
Pileated Woodpecker	8"	8"	16-24"	12-20"	3"x4"	15-20 ft.
Red-headed Woodpecker	6"	6"	12-15"	9-12"	2"	10-20 ft.
Yellow-bellied Sapsucker	5"	5"	12-15"	9-12"	1-1/2"	10-20 ft.
House Wren	4"	4"	6-8"	4-6"	1-1/4"	5-10 ft.
Barn Owl	10"	18"	15-18"	4"	6"	12-18 ft.
Screech Owl, Kestrel	8"	8"	12-15"	9-12"	3"	10-30 ft.

Sources: 113, 112, 45

TABLE 5-1

GRAINS, GRASSES AND LEGUMES FOR USE IN FOOD PLOTS AND OTHER AREAS

GRAIN	WILDLIFE ATTRACTED	Season Used Most
Barley	Foliage: Geese, black ducks; grain: Hungarian partridge, pheasants, crows	fall
Browntop Millet	Ducks, mourning doves	fall
Buckwheat	Mourning doves, Hungarian partridge, deer, songbirds, ducks	fall
Corn	More than 100 species, including pheasants, quail, Hungarian partridge, mourning doves, ducks, geese, meadowlarks, blue jays, raccoons (unless sweet corn is available), foxes, squirrels, many songbirds; also provides cover	fall, esp. winter (stands tall above snow)
Oats	Hungarian partridge, prairie chickens, pheasants, wild turkeys, many songbirds	fall
Rye	Foliage: cottontail rabbits, deer, geese; seeds: pheasants, quail	spring
Sorghum	Quail, wild turkeys, ducks, many songbirds	fall
Wheat	Food for 94 species; winter wheat foliage: geese, cottontail rabbits, deer; grain: mourning doves, Hungarian partridge, many songbirds, pheasants, quail	summer
GRASSES	**WILDLIFE ATTRACTED**	**Season Used Most**
Kentucky Bluegrass	Nesting cover: pheasants, quail, rabbits; food: rabbits, geese	winter, fall
Smooth Brome	Nesting cover: pheasants, quail, rabbits, songbirds; food: deer, rabbits	winter, fall, spring
Reed Canary Grass	CAUTION: No longer recommended—invasive	

GRASSES	WILDLIFE ATTRACTED	Season Used Most
Orchardgrass	Nesting cover: game birds, rabbits, songbirds	late spring, summer
Redtop	Nesting cover: game birds, rabbits, songbirds	spring
Switchgrass	Nesting cover: pheasants, quail, rabbits, songbirds; food: songbirds, quail, rabbits	late spring, summer
Timothy	Nesting cover: game birds, songbirds, rabbits; food: rabbits, deer, songbirds	late spring, summer, fall
LEGUMES	**WILDLIFE ATTRACTED**	**Season Used Most**
Alfalfa	Nesting cover: game birds, rabbits; food: rabbits, deer	late spring, summer, fall
Birdsfoot Trefoil	Food: ruffed grouse, deer, rabbits	late spring, summer
Clovers (Alsike, White Ladino, Red)	Nesting cover; foliage: food for ruffed grouse, wild turkeys, woodchucks, rabbits, deer	late spring, summer
Cowpeas	Seeds: mourning doves, quail	fall
Crownvetch	Cover: rabbits	summer, fall
Korean Lespedeza	Food: quail, deer	fall
Lespedeza, Sericea	Cover: rabbits, quail	summer, fall
Partridge Pea	Food: quail	fall
Soybean	Seeds: mourning doves, quail, pheasants; foliage: rabbits	summer, fall
Sweet Clovers	CAUTION: These grow faster than native plants, shading and inhibiting growth. Some are regarded as plants to remove, not encourage	

WEEDS	WILDLIFE ATTRACTED	Season Used Most
Crabgrass	Wild turkey, junco, chipping sparrow, field sparrow, savannah sparrow, tree sparrow	
Panic Grass	American tree sparrow, field sparrow, white-throated sparrow	summer
Bristle Grass	Mourning dove, Hungarian partridge, red-winged blackbird, bobolink, cowbird, dickcissel, junco, horned lark, meadowlark, various sparrows	fall
Sedges	Sora rails, snow buntings, swamp sparrows	summer, spring
Smartweed	Black & mallard ducks, blue-winged teal, quail, cardinals, common redpolls, fox, song, and white-throated sparrows	
Ragweed	Hungarian partridge, quail, goldfinches, juncos, horned larks, common redpolls, fox, harris, lincoln, song, vesper, white-crowned and white-throated sparrows	
Dropseed Grass	Juncos	fairly good in spring, summer
WARM SEASON GRASS	**WILDLIFE ATTRACTED**	**Season Used Most**
Big Bluestem	Bobolinks, eastern and western meadowlark, red-winged blackbird, savannah sparrow, grasshopper sparrow, Henslow's sparrow, song sparrow, dickcissel, upland sandpiper, pheasants, gray (Hungarian) partridge	late spring, summer, fall
Little Bluestem	Songbirds	late spring, summer
Sideoats Grama	Songbirds	late spring, summer
Indian Grass	Songbirds, pheasants, gray (Hungarian) partridge	late spring, summer

Sources: 118, 14, 72, 89

TABLE 6-1

HOMEMADE AND COMMERCIAL DEER REPELLENTS

Contents	Brand Name	Uses	Estimated Effectiveness	Durability	Effectiveness of Renewed Application
Feather Meal	—	In 2+ cloth bags on woody plants	90-95%	30-90 days	Same
Meat Meal	—	In 2+ cloth bags on woody plants	90-95%	30-90 days	Same
Meat Meal/pepper	Greenscreen	In 2+ cloth bags on woody plants	95-100%	30-90 days	Same
Blood Meal	—	Apply to area to be protected	90-100%	3-10 days (washes off with rain)	Same
Soap Bars	—	2+ bars on woody plants	80-90%	30-90 days	Same
Liquefied Eggs In Water	—	Spray on any plant	80-90%	3-7 days	Same
Putrescent Whole Egg Solids	Deer-Away	Spray or dust on ornamental and non-bearing fruit trees	95-100%	21-42 days	Same
Ammonium Hydroxide	Hinder	Spray on any plants	80-95%	7-14 days (washes off with rain)	Same
Capsaicin	Hot Sauce	Spray on ornamental and non-bearing fruit trees	0-50%	15-30 days	Same or less
Thiram	Bonide, Lesco, Spotrete	Spray on ornamental and non-bearing fruit trees	50-75%	15-30 days (90 days with sticker)	Same

Benzyl Diethyl Ammonium Saccharide Thymol	Ro-Pel	Spray on ornamental and non-bearing fruit trees	0-50%	7-14 days	Same or less
Contents	**Brand Name**	**Uses**	**Estimated Effectiveness**	**Durability**	**Effectiveness of Renewed Application**
Denathonum Benziaata (Bittrex)	Tree Guard	Spray on ornamental and non-bearing fruit trees	50-75%	30-60 days	Same
Garlic	—	Spray on ornamental and non-bearing fruit trees	90-100%	30-60 days	Same
Mixtures Of Above	*Deerbuster, Deerstopper	Spray on ornamental and non-bearing fruit trees	95-100%	30-60 days	Same
Cat Urine & Feces (Lion Urine, Feces)	—	Apply to area to be protected	50-75%	7-14 days	None
Moth Balls	—	Apply to area to be protected	0-50%	3-14 days	None
Human Hair	—	In 2+ cloth bags on woody plants or spread on ground around plants	0-50%	3-7 days	Variable

Information provided by Deer Barriers . . . fencing, repellents, dog restraint systems, scare devices, Michigan State University Extension Bulletin E-2672, August, 1998.